SpringerBriefs in Applied Sciences and Technology

Series editor

Janusz Kacprzyk, Polish Academy of Sciences, Systems Research Institute, Warsaw, Poland

SpringerBriefs present concise summaries of cutting-edge research and practical applications across a wide spectrum of fields. Featuring compact volumes of 50–125 pages, the series covers a range of content from professional to academic.

Typical publications can be:

- A timely report of state-of-the art methods
- An introduction to or a manual for the application of mathematical or computer techniques
- A bridge between new research results, as published in journal articles
- A snapshot of a hot or emerging topic
- An in-depth case study
- A presentation of core concepts that students must understand in order to make independent contributions

SpringerBriefs are characterized by fast, global electronic dissemination, standard publishing contracts, standardized manuscript preparation and formatting guidelines, and expedited production schedules.

On the one hand, **SpringerBriefs in Applied Sciences and Technology** are devoted to the publication of fundamentals and applications within the different classical engineering disciplines as well as in interdisciplinary fields that recently emerged between these areas. On the other hand, as the boundary separating fundamental research and applied technology is more and more dissolving, this series is particularly open to trans-disciplinary topics between fundamental science and engineering.

Indexed by EI-Compendex and Springerlink.

More information about this series at http://www.springer.com/series/8884

Flavio Pendolino · Nerina Armata

Graphene Oxide in Environmental Remediation Process

 Springer

Flavio Pendolino
Padova
Italy

Nerina Armata
Palermo
Italy

ISSN 2191-530X   ISSN 2191-5318   (electronic)
SpringerBriefs in Applied Sciences and Technology
ISBN 978-3-319-60428-2   ISBN 978-3-319-60429-9   (eBook)
DOI 10.1007/978-3-319-60429-9

Library of Congress Control Number: 2017944297

Printed on acid-free paper

This Springer imprint is published by Springer Nature
The registered company is Springer International Publishing AG
The registered company address is: Gewerbestrasse 11, 6330 Cham, Switzerland

Preface

Nanomaterials exhibit more remarkable properties than other conventional materials in numerous areas of interest including the environmental remediation. Graphene oxide is a smart nanomaterial which can be employed in this field due to its physical chemical properties. In the last decade, the possible application of graphene oxide in the production of nanocomposites, for example, in electronic devices, or starting materials for pristine graphene has been objected of discussion. Recently, graphene oxide is figured out as a promising material in environmental applications. This book covers the current investigations on the implementation of graphene oxide as a sorbent material for removal technology. Chapter 1 begins with the state of the art of graphene oxide for novel adsorbent materials. Chapter 2 briefly illustrates the methods for wet synthesis of graphene oxide. The properties of this material are showed in the second part of this chapter describing the spectroscopic and topographic characteristics. In the last part, the models of the structure of graphene oxide are reported. Chapter 3 covers the environmental aspects from the point of view of regulation with particular attention for European area. Chapter 4 summarizes the latest founds on removal of contaminants from water systems. Chapter 5 addresses the conclusion and future potential use of graphene oxide in advanced removal technology.

The authors thank Prof. P. Colombo (University of Padova, Italy), Prof. A. Maddalena, Dr. AC. Nale (University of Padova, Italy), Dr. T. Masullo (CNR, Italy), Dr. A. Cuttitta (CNR, Italy), Dr. F. Busolo (Ordine dei Chimici) for collaboration on the graphene oxide and related projects. We also deeply thank Dr. Mayra Castro in Springer for inviting us to write this book and for many helpful advices during the preparation of this book.

Padova, Italy   Flavio Pendolino
Palermo, Italy   Nerina Armata
April 2017

Contents

Chapter 1
Introduction

Abstract Since the water crisis is facing, purification methods and removal technology are the object of discussion. Carbon-based materials are largely employed as filters and in particular graphene oxide is considered for its exceptional physical chemical properties as a good candidate for the fabrication of innovative devices.

Keywords Wastewater · Water purification · Adsorbent materials · Carbon-based materials · Graphene oxide

Water has an essential role in life and is at the core of natural ecosystems and climate regulation. Hence, water resources must be carefully managed to ensure an adequate quantity and quality to the population and ecosystem. Water scarcity and pollution are a threaten to human health and quality of life and the World Bank has estimated a worldwide usage of fresh water of 3.9 billion/m3 comprising of domestic (18%), agriculture (60%) and industrial (22%) wastewater in 2014 (latest update data) [18]. In Europe, the water issues afflict 17% of the European territory and it has been predicted that a million people will experience a severe water stress by 2050 [3, 6, 10, 13, 16]. So far, we are facing a water crisis.

Conventionally, water purification methods include chemical, biological degradation, disinfection (oxidation/ozonation), carbon adsorption and filtration/removal processes [2, 14]. The removal technology is an essential approach for water treatments, which are based on the physical chemical properties of an adsorbent substrate. Adsorbent materials benefit from low-cost production, reduced hazardous character for human health and have the ability to sequester a wide range of emerging contaminants, as well as, nanomaterials, heavy metals and organic molecules. Several materials are currently employed in filtration technology such as polymers, high surface area materials (zeolites, clay), wood, carbon-based substrates [9]. Moreover, development of nanotechnologies and smart materials, possessing unconventional properties, fosters the use of innovative solutions for water treatment. The combination of removal processes and smart materials is a promising methodology and technology for improving the quality of the water and its reuse.

Carbon-based materials are largely implemented in numerous technological applications due to their low-cost of production. Typically, activated carbon with a high

© The Author(s) 2017

F. Pendolino and N. Armata, *Graphene Oxide in Environmental Remediation Process*, SpringerBriefs in Applied Sciences and Technology, DOI 10.1007/978-3-319-60429-9_1

surface area is commonly used in filter technology since decades. Among novel nano-materials, graphene and its oxidized form is the most rising alternative in removal technologies. They are extensively investigated materials with a growing number of related scientific publications [4, 5, 7]. Despite that, water reuse/recycle has been marginally investigated for environmental remediation applications [8, 11, 19]. In fact, the majority of scientific literature reports on the interaction between graphene (oxide) and a target molecule with emphasis only on physical or chemical properties. Among graphene based materials, graphene oxide (GO) was selected as a suitable substrate, due to its exceptional physical chemical properties. Graphene oxide is a single layer of carbon atoms arranged in a honeycomb lattice and containing oxygen functionalities. In general, graphene oxide allows the adsorption of both organic and inorganic molecules through its basal plane and hydroxyl and carbonyl groups [15]. Recently, Liu and coworkers reported on fast dyes removal from water by means of a filter based on graphene oxide [12] and Aba and coworkers showed a significant increase of high-efficiency in liquid filtration by means of graphene oxide [1]. Wang and coworkers and Li and coworkers collected the most updated information about the use of graphene oxide for prototype pollutions and applications which range from removal of metals (Cu, Pb, Cr, etc.) to dyes and gas [11, 17]. All of these studies endorse the concept that graphene oxide can be employed as an advanced removal agent for molecules, ions and metal atoms in a polar medium. Moreover, little importance is given to the quality of the used graphene oxide, notwithstanding the fact that this parameter strongly affects the reactivity. In fact, a different form/structure of GO (liquid phase, 3D structure, the amount of oxygen, etc.) interacts with a contaminant using diverse reaction paths with the consequence that the removal efficiency is affected. The GO is currently a good potential material to remove pollutions, especially because of ambivalent hydrophobic/hydrophilic character.

References

1. Aba NFD, Chong JY, Wang B, Mattevi C, Li K (2015) Graphene oxide membranes on ceramic hollow fibers—microstructural stability and nanofiltration performance. J Membr Sci 484:87–94
2. Baláž M (2014) Eggshell membrane biomaterial as a platform for applications in materials science. Acta Biomater 10(9):3827–3843
3. Bixio D, Thoeye C, De Koning J, Joksimovic D, Savic D, Wintgens T, Melin T (2006) Wastewater reuse in Europe. Desalination 187(1–3):89–101
4. Dreyer DR, Todd AD, Bielawski CW (2014) Harnessing the chemistry of graphene oxide. Chem Soc Rev 43(15):5288–5301
5. Eigler S, Hirsch A (2014) Chemistry with graphene and graphene oxide-challenges for synthetic chemists. Angew Chem Int Ed 53(30):7720–7738
6. Eurepean Commission (2014) Water reuse in Europe. Technical report. Joint Research Centre (JRC)
7. Gambhir S, Jalili R, Officer DL, Wallace GG (2015) Chemically converted graphene: scalable chemistries to enable processing and fabrication. Nature 7(6):e186
8. Georgakilas V (2014) Functionalization of graphene. Wiley, New York

9. Hu A, Apblett A (eds) (2014) Nanotechnology for water treatment and purification. Springer, Heidelberg
10. Iranpour R, Cox H, Kearney RJ (2004) Regulations for biosolids land application in US and European Union. J Residuals Sci Technol 1:209–222
11. Li F, Jiang X, Zhao J, Zhang S (2015) Graphene oxide: a promising nanomaterial for energy and environmental applications. Nano Energy 16:488–515
12. Liu F, Chung S, Oh G, Seo TS (2012) Three-dimensional graphene oxide nanostructure for fast and efficient water-soluble dye removal. ACS Appl Mater Interfaces 4(2):922–927
13. Norton-Brandão D, Scherrenberg SM, van Lier JB (2013) Reclamation of used urban waters for irrigation purposes—a review of treatment technologies. J Environ Manage 122:85–98
14. Smith SC, Rodrigues DF (2015) Carbon-based nanomaterials for removal of chemical and biological contaminants from water: a review of mechanisms and applications. Carbon 91:122–143
15. Upadhyay RK, Soin N, Roy SS (2014) Role of graphene/metal oxide composites as photocatalysts, adsorbents and disinfectants in water treatment: a review. RSC Adv 4(8):3823–3851
16. Vajnhandl S, Valh JV (2014) The status of water reuse in European textile sector. J Environ Manage 141:29–35
17. Wang S, Sun H, Ang HM, Tadé MO (2013) Adsorptive remediation of environmental pollutants using novel graphene-based nanomaterials. Chem Eng J 226:336–347
18. World Bank (2017). Annual freshwater withdrawals. www.data.worldbank.org/. Accessed 30 Apr 2017
19. Zhao J, Liu L, Li F (2015) Graphene oxide: physics and applications. Springer Briefs in Physics. Springer, Heidelberg

Chapter 2
Synthesis, Characterization and Models of Graphene Oxide

Abstract The chapter gives an overview of the most important synthetic methods to produce graphene oxide via wet chemistry. Brodie-Staudenmaier-Hummers and free-water based approaches are reported and where it is considered the influence of synthesis on the graphene oxide structure. Physical chemical and morphological characterisations are described in the second part of the chapter. In the last section, details of theoretical calculation and modelling of graphene oxide structures are presented.

Keywords Graphene oxide synthesis · Physical chemical characterization · Morphology · Molecular modelling

Graphene oxide is not a natural product and results in a non-stoichiometric "molecule". Basically, *graphene oxide is defined as a layer of graphene decorated with oxygen functionalities*, such as hydroxyl (OH), carbonyl (C=O) and alkoxy (C–O–C) groups. Practically, graphene oxide is made when a pristine graphene is oxidized, following a general master equation

$$\text{Carbon Layer} \xrightarrow{Oxidation} \text{GO} \qquad (2.1)$$

In Fig. 2.1a–c is shown a typically representation of GO structure, a yellow-amber dispersion and an oven dry solid of graphene oxide, respectively. The description of a single unit of GO mainly depends on the size of the graphene basal plane and the number of oxygen functionalities. The size of GO is controlled by the carbon source, often graphite. Besides, the oxygen domains are subjected by the synthetic procedure. Several grades of oxidation can be found relying on the number of oxygens. The diagram in Fig. 2.1b shows that graphene oxide exists in a range 10–50 at.%O limited by temperature of about 60–70 °C. In this oxidation range, pristine graphene oxide is observed with different solubility and it roughly splits into two subgroups that we name as *low solubility GO* (15–35 °C) and *high solubility GO* (35–50 °C), which depends on the oxidation grade. At zero oxidation (0 at.%O), *pristine graphene* is found, while a reduced graphene oxide (rGO) can be observed with a oxidation up to 10 at.%O. The maximum theoretical value of oxidation for graphene cannot be over

© The Author(s) 2017
F. Pendolino and N. Armata, *Graphene Oxide in Environmental Remediation Process*, SpringerBriefs in Applied Sciences and Technology, DOI 10.1007/978-3-319-60429-9_2

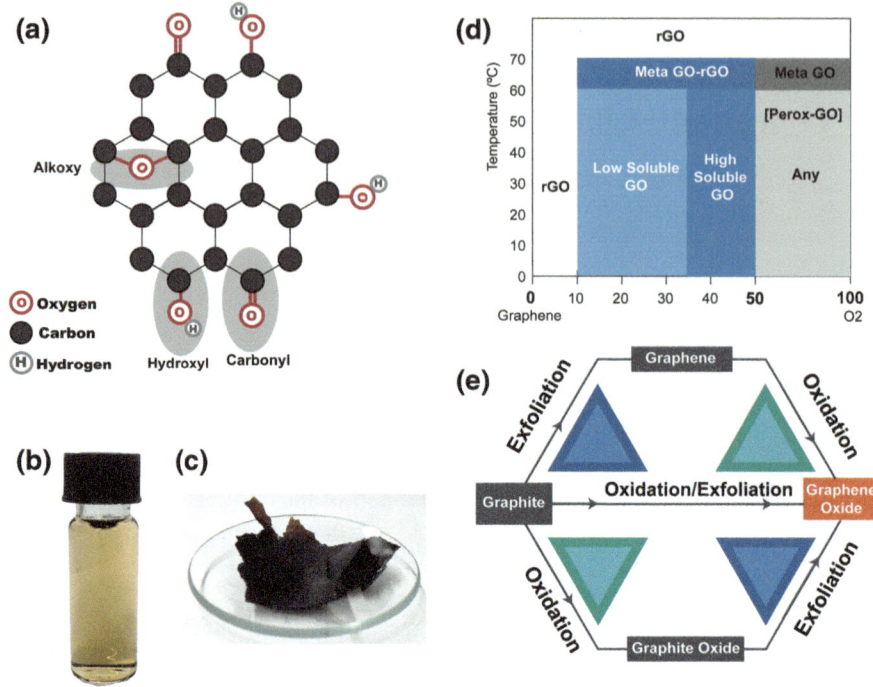

Fig. 2.1 **a** Draw of graphene oxide structure. The illustration shows the hexagonal arrangement of carbon atoms and the presence of oxygen domains. **b** A drafted diagram of the oxidation degree (%O) of graphene oxide versus temperature. Below 10 at.%O and above 70 °C reduced graphene oxide is found, while over 50 at.%O a GO structure is not clear. The rage between 10 and 50 at.%O shows the presence of GO possessing various degree of solubility. A meta state can be found around 60–70 °C. **c** Digital photograph of 0.05 mg/mL aqueous dispersion and **d** solid graphene oxide. **e** Scheme of GO production starting from graphite as carbon source. Three paths can be described to get as final product the graphene oxide material

50 at.%O because of the sp^2 hybridization of carbon atom in the graphene hexagons. Eventually, an oxidation over the 50 at.%O may occur with a fully oxidation by peroxide groups (O_2^{2-}), but this oxidation grade is not yet observed.

2.1 Synthetic Methods for Graphene Oxide

Graphene oxide can be synthetized in a *dry* or *wet* medium. The dry synthetic approach consists in an oxidation reaction of graphene through atomic oxygen in ultrahigh vacuum conditions [21, 47], to exposure to molecular oxygen [44, 49] and treating with ozone under ultraviolet light [10, 22]. By contrast, a suitable approach consists of wet synthesis in which *graphite* is typically used as graphene source, due to its natural abundance and inexpensive cost. In Fig. 2.1e, the three main reaction routes are shown. The first approach starts with using graphene, produced with mechanical method, followed by a further oxidation, while the second one is based on exfoliation in aqueous media by means of ultrasonic treatment [4, 43]. The last route takes in a concurrent of oxidation and exfoliation process in strong acidic medium as found in Brodie-Staudenmaier-Hummers method. *Overall, these paths bring to the formation of graphene oxide, but the structural properties of each type of GO are different, such as the structure or the reactivity sites.*

In the following paragraphs, in chronological order, are reported approaches of preparing graphene oxide. During the last century, several methods were proposed and the three major methods are Brodie [5], Staudenmaier [45] and Hummers [23]. From these basic methods a number of variations were derived to improve the overall yield and quality of the product, for instance the Tour method [31]. Recently, a "primitive" approach was adopted by a number of researchers which consists in a free-water exfoliation and oxidation of graphite through strong oxidizing agent in a strong protic medium (often H_2SO_4) [35, 46].

2.1.1 Brodie-Staudenmaier-Hummers Based Methods

Brodie Method. The first documented synthesis of graphitic oxide material is attributed to Brodie in 1859 [5]. His study was focused on finding the weight of graphite and, as common at that time, a series of chemical reactions were investigated to elucidate the properties of a novel material. Thus, graphite was mixed with potassium chloride ($KClO_3$) and solubilized in fumic nitric acid to oxide the sample and inferring the molecular weight. Further oxidation processes were carried out on the sample up to any changes were visible. The elemental analysis revealed a composition of *circa* 60% C, 2% H and 38% O. The resulting product, a mixture of graphene and graphite oxide, was soluble in pure water.

Staudenmaier Method. In 1898 Staudenmaier [45] improved the Brodie's reaction by adding sulfuric acid, to increase the acidity of the mixture, and several aliquots of solid $KClO_3$, over the course of reaction. Despite that, Brodie and Staudenmaier method generates ClO_2 toxic gas which rapidly decomposes in air giving explosions. These modifications led to a more oxidized graphitic material and a simplification of the reaction.

Hummers Method. In 1958, Hummers and Offeman [23] proposed an alternative way to oxide graphite improving the safe operational conditions with a drastic reducing time, from 10 to 2 days. They mixed graphite with concentrated sulfuric acid (H_2SO_4), sodium nitrate ($NaNO_3$) and potassium permanganate ($KMnO_4$) to obtain a brownish grey pasty. The suspension was diluted with water and hydrogen peroxide (H_2O_2) was added to get a higher oxidation degree and to eliminate manganese from the dispersion (yellow-brown mixture). Finally, the sample was filtered and washed with warm water. They achieved the same degree of oxidation reported by Staudenmaier, however, the amount of GO results very little. The weakness of this

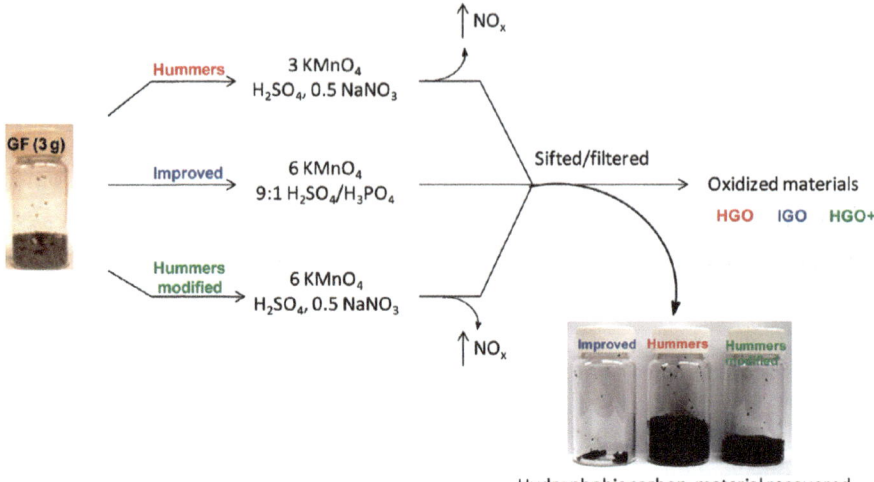

Fig. 2.2 Procedure scheme for Tour method. The starting material is expanded graphite to be used for producing graphene oxide through Tour method and in comparison with Hummers and its modification. Reprinted with permission from [31]. Copyright (2010) American Chemical Society

method includes a time-consuming of the separation and purification process. From Hummers method a huge number of variation/optimization approaches has been developed and a typical GO product is made by flakes of about 1 nm and a lateral size of 1 μm.

Tour Method. An improvement of Hummers. method was proposed by Tour's group at Rice University in 2010 [31]. They have substituted the sodium nitride with phosphoric acid in a mixture of H_2SO_4/H_3PO_4 (9:1) and increasing the amount of $KMnO_4$. The advantage of this method consists in no generation of toxic gases, such as NO_2, N_2O_4 or ClO_2, in the reaction and an easy temperature control. The authors claim that the presence of phosphoric acid generates a more intact graphitic basal plane. A comparison of the improved method with the conventional and modified Hummers's procedures can be seen in Fig. 2.2. The advantage of the Tour method consists in a production of graphene oxide having a higher hydrophilic degree, in contrast with GO produced by Hummers method (see Table 2.1). Hence, this graphene oxide results more oxidized and soluble.

2.1.2 Free-Water Oxidation Method

The Free-Water Oxidation method takes advantage of a reaction between expanded graphite and oxidizing agent in a free-water medium. Because of inorganic carbon is essentially inert at room temperature, its solubilization/dispersion in a solvent

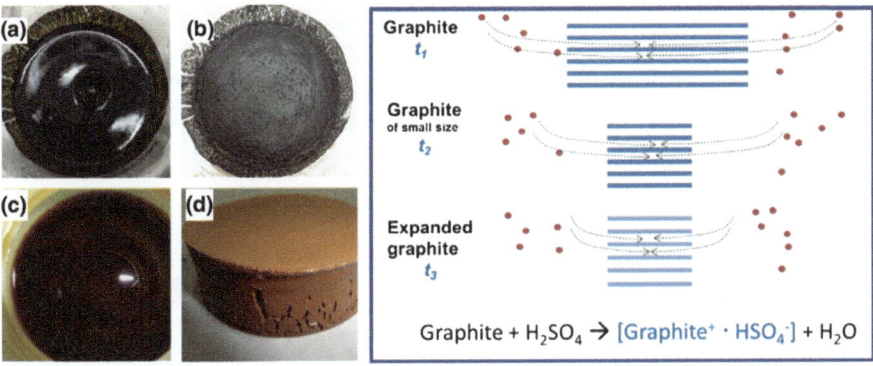

Fig. 2.3 (*Left*) Reaction scheme for Sun method. (*Right*) Photos of **a** mixture of reagent, **b** volumetric expansion (foam-like), **c** hydrolysis and **d** concentrated dispersion after purification. Reprinted from [46], Copyright (2013), with permission from Elsevier

needs strong protic acid or mixture of warm acids, such as sulphuric or nitric acid. Moreover, a strong oxidizing agent, as potassium permanganate ($KMnO_4$), ensures the bonds of oxygen functionalities to inorganic carbons. From historical reason, the free-water oxidation methods derived from a modification of the Hummers method in which some limitations, such as hazardous reagents, were improved.

Sun Method. In 2013, Sun and Fugetsu at Hokkaido University [46] introduced a more direct method to produce graphene oxide. They used expanded graphite as carbon precursor. The potassium permanganate had twofold effects: intercalating agent and oxidizing agent. The intercalation of $KMnO_4$ between graphitic layers produced a further *spontaneous expansion* which looks like a foam of graphitic material, as shown in Fig. 2.3b. The reaction occurs in acidic medium of sulfuric acid. The ratio Graphite:H_2SO_4 was reduced to 1:20 and additional reagents were eliminated from the reaction procedure. For this reason, Sun protocol can be considered as one of the first green procedure of among the wet synthetic methods.

Peng Method. Very recently (2015), Peng and co-workers [36] proposed a scalable and green method (Fig. 2.4) to produce graphene oxide, using potassium ferrate (K_2FeO_4) as strong oxidant. This compound avoids the introduction of heavy metals or the formation of toxic gases during the preparation. In this method, a mixture of graphitic flakes and K_2FeO_4 dispersed in concentrated sulphuric acid, were loaded into a reactor and stirred for 1h at room temperature. The product was water-washing by repeated centrifugation to obtain highly water soluble graphene oxide.

4-Steps Method. This method derived from the basic exfoliation-oxidation procedure and was improved by Pendolino and co-workers [35]. The method consists of 4 reaction steps controlled by temperature, which affects strongly the final product. A scheme of reactions is shown in Fig. 2.5 in which two paths are proposed depending on temperature. In the first step the oxidation process for the mixture graphite-$KMnO_4$ dispersed in concentrated sulfuric acid gets a pasty slurry. The sec-

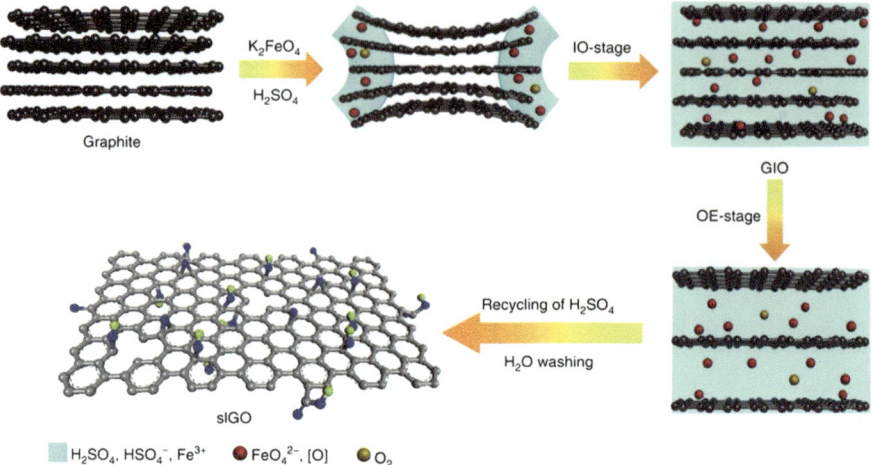

Fig. 2.4 Reaction mechanism proposed by Peng and co-workers for the synthesis of graphene oxide using K_2FeO_4 as oxidizing agent. Reprinted from [36]. This work is licensed under a Creative Commons Attribution 4.0 International License

ond step (warm) consists of the exfoliation of graphite and is the most critical one. In fact, the production of *graphene oxide* is limited by temperature and only occurs when the water bath is around 30 °C. By contrast, the exfoliation is suppressed for lower temperature with a resulting *graphite oxide* (cold). The hydrolysis, at 90 °C for 1 h, completes the third step. The purification of product is performed by centrifugation using warm water up to the neutrality of the dispersion, at the forth step. Through the 4-Steps method two different products can be synthetized just controlling the temperature along the reaction. The advantage of this method is related to the improved of safety operational conditions (limiting explosive reaction due to Mn_2O_7 in concentrated sulfuric acidic for temperature above 55 °C) and the production of a type of GO which contains an amount of oxygen domains lower to about 20–30 at.%O. This type of GO can be employed for filter/remediation or biosystems due to the low toxic effect.

In all the above synthetic methods to prepare graphene oxide some limitations are encountered. Primary, lab safety ascribed to hazardous reagents [12, 13, 19]. The use of sodium nitrate or potassium chlorate in Brodie or Staudenmaier methods results explosive, while sodium nitrate (Hummers) or fuming nitric acid, introduce heteroatoms or defects on the GO structure that affects the reactivity [9, 11]. Next, the presence of dimanganese heptoxide (Mn_2O_7) in a solution of sulphuric acid detonates with a temperature over 55 °C. An other important factor is the quality and the grain size of graphite. In fact, defect free structure gives higher quality of graphene oxide, as reported by Chen et al. [8] and the size of grain establishes the size for the graphene basal plane. Moreover, the ratio C/O ranges from 0.7 (∼55 at.%O) to 3.5 (∼20 at.%O) relying on the synthetic method, as well as, the resistivity

Fig. 2.5 Reaction scheme for the 4-Steps method. The temperature influences the final product according with the "warm" or "cold" path. The graphene oxide is only obtained when the warm path is followed, and, by contrast, graphite oxide is produced at lower temperature below 30 °C

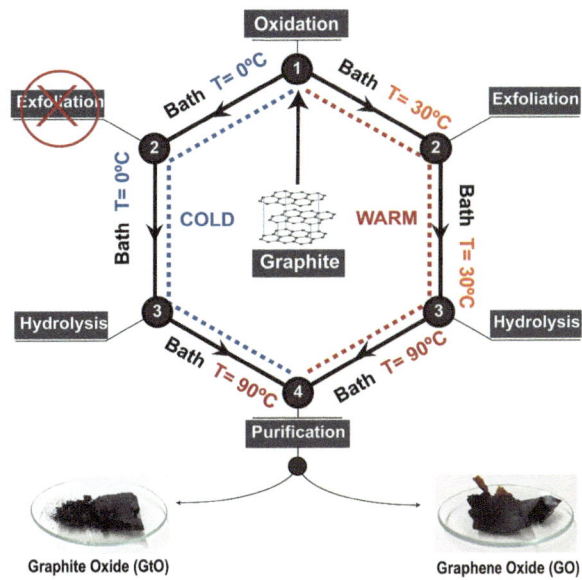

Table 2.1 Summary of the main synthetic methods used to prepare GO

Method	Oxidant	Solvent	Additive	C/O	Resistivity[a]	Ref.
Brodie	$KClO_3$	HNO_3	–	2.4–2.9	0.15–60	[5, 39–41, 51]
Staudenmaier	$KClO_3$	Fuming HNO_3	–	2.2	120	[30, 40, 45]
Hummers	$KMnO_4$	H_2SO_4	$NaNO_3$	1.8–2.5	0.005–0.01	[23, 25, 40, 43, 51]
Tour	$KMnO_4$	H_2SO_4	H_3PO_4	0.7–1.3	0.2–1000	[18, 26, 31]
Sun	$KMnO_4$	H_2SO_4	–	2.5	0.18	[46]
Peng	K_2FeMO_4	H_2SO_4	–	2.2	2.7	[36]
4-Steps	$KMnO_4$	H_2SO_4	–	3.5	23	[35]

[a] 10^5 $\Omega \cdot m$

varies from 0.005 to 1000 × 10^{-5} $\Omega \cdot m$, as shown in Table 2.1. Details on synthesis and functionalization of graphene oxide can be consulted at [6, 16, 17, 53].

Furthermore, the purification is overlong and time-consuming. The filtration using paper filter is not a practicable approach because of the unpurified mixture possesses a colloidal behaviour. Dialysis or centrifugation required a large number of washing cycles. All these aspects strongly affect the final product. To conclude, there is not an exhaustive method or procedure for producing a "standard" graphene oxide, because of for each synthetic method a *different type of graphene oxide* is produced. Hence, graphene oxide exhibits different physical chemical properties, such as structure and reactivity. The discrepancy between the structure and reactivity of graphene oxide is in the mosto of the cases due to the synthetic method and the carbon source, as reported in literature. The standardization of synthesis appears one of the major obstacle in using GO for advanced applications. Nervertheless, different type of graphene oxide, produced with different methods, can enlarge the implementation range of this material by *modulating* its properties and opening new operating routes. Therefore, different type of GO can be considered an advantage for a future use of this nanomaterials.

2.2 Characterizations

The physical chemical characterization of graphene oxide is not often of easy interpretation because this material is made of carbons and oxygens as the majority of organic molecules. Thus, the peculiar spectroscopic signals are blended by conventional carbon and oxygen signals. In the following paragraphs, the most regular techniques and the representative interpretation to identify the GO are shown.

FTIR. The FTIR is an efficient tool for a rapid characterization of GO. It becomes common to interpret FTIR signals as referred to hydroxyl (OH), epoxy (C–O–C) and ketone (C=O) groups. The corresponding vibrational frequencies for the stretching mode are around $3400 \, cm^{-1}$, $1090 \, cm^{-1}$ and a doublet at $1700 \, cm^{-1}$, respectively. A typical FTIR spectrum is reported in Fig. 2.6a. It is obvious that the fingerprint region of the IR spectrum of GO varies. The plots a and b show a spectrum of two different synthesis of GO having a oxygen content of 20 at.%O, while plot c represents a commercial GO possessing 50 at.%O. The most of the difference are found for peaks at $1090 \, cm^{-1}$ and the first peak of the doublet around $1700 \, cm^{-1}$. An other example of FTIR spectrum is reported by Bagri et al. [3]. They obtained the following vibration frequencies: hydroxyl $3050–3800 \, cm^{-1}$, carbonyls $1750–1850 \, cm^{-1}$, carboxyls $1650–1750 \, cm^{-1}$, C=C $1500–1600 \, cm^{-1}$ and ethers/epoxides $1000–1280 \, cm^{-1}$. A different interpretation for the two peaks around $1700 \, cm^{-1}$ is reported by Pendolino et al. [33] for GO spectrum having an oxygen content of about 20–30 at.%. Similar frequencies are reported for pristine GO from water dispersion, bands a centred at hydroxyl $3410 \, cm^{-1}$, carbonyls 1730 and $1620 \, cm^{-1}$, and ethers/epoxides $1090 \, cm^{-1}$. Authors argue that these peaks ($1730 \, cm^{-1}$, $1620 \, cm^{-1}$) are not different carbonyl/carboxyl groups but the doublet are interpreted as a *keto-enol tautomers*, an equilibrium that occurs in molecules carrying keto and enol groups. As reported previously [34], the tautomeric equilibrium is shifted toward the enol form when the dielectric constant is increased, while has an intensity ratio 1:1 in water dispersion. This fact makes evidence of the interpretation for keto-enol tautomerism.

Raman. Raman spectrum for the GO shows only two broaden peaks which represent the G and D band (Fig. 2.6b). The first band is found around $1580 \, cm^{-1}$ and is attributed to the in-phase vibrations of ordered crystal structure while the D band

Fig. 2.6 a FTIR spectra for GO. The profile a GO with a oxygen content of 20 at.%O following the 4-step synthetic method, profile b different synthesis of GO with a oxygen content of 20 at.%O following the 4-step synthetic method, profile c commercial GO with a oxygen content of 50 at.%O. **b** Raman spectra for graphite, GO and reduced GO by γ ray. Reproduced from Ref. [52] with permission from the Royal Society of Chemistry

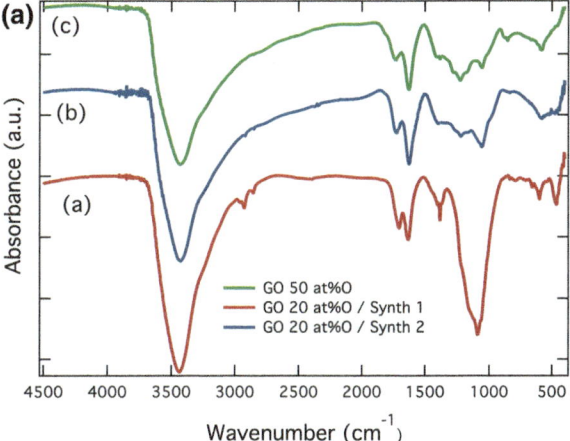

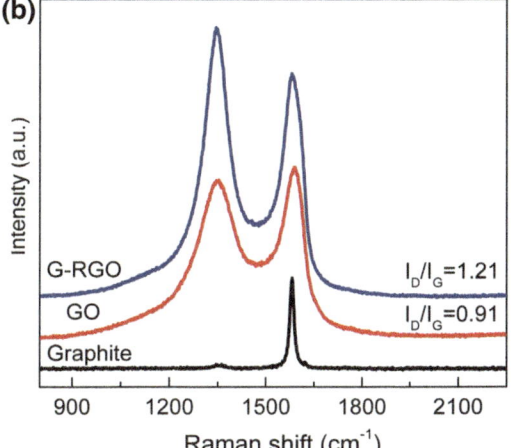

($\sim$1350 cm^{-1}) is assigned to the disorder crystal structure. This fact correlates the G band to the sp^2 carbon and the D band to the presence of sp^3, hence, to oxygen domains [28, 52].

UV-vis. The UV-vis spectra for GO shows one adsorption signal followed by a shoulder in the range 190–900 nm. The region 400–900 nm is not affected by any absorptions. The maximum absorbance is found around $\lambda = 230$ nm which is attributed to $\pi \rightarrow \pi^\star$ transitions in conjugated systems. The shoulder occurs at around $\lambda = 300$ nm and is often assigned to the $n \rightarrow \pi^\star$ transitions of a carbonyl group. The optical absorption is changing with the number of layers, as reported by Lai et al. [29]. They observed a disappearance of maximum adsorption band towards a lower wavelengths of 196 nm for thick-layer >10 layers (Fig. 2.7).

Fig. 2.7 **a** UV-vis spectra for GO dispersions at different sonication time and **b** histograms vs thickness. Reproduced from Ref. [29] distributed under a Creative Commons Attribution 3.0 Unported License

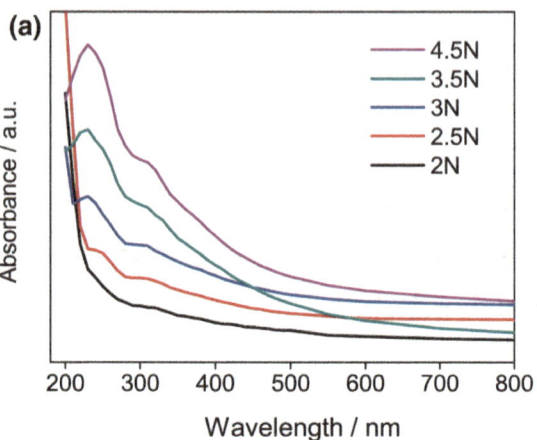

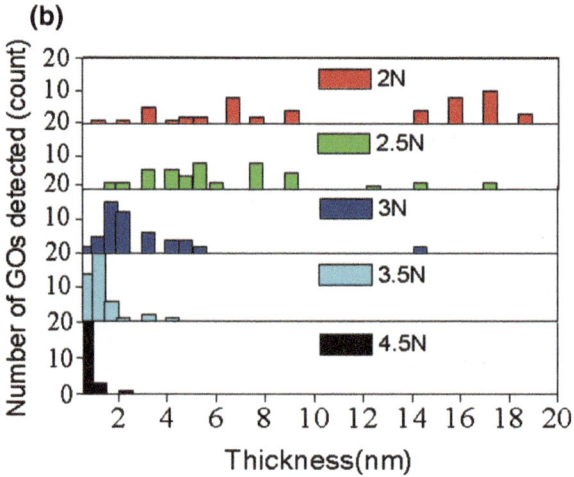

Fig. 2.8 XRD patterns from GO dispersion in DMF, CHP, THF, Acetone, water and ethanol solvent. Reprinted with permission from [24]. Copyright (2013) American Chemical Society

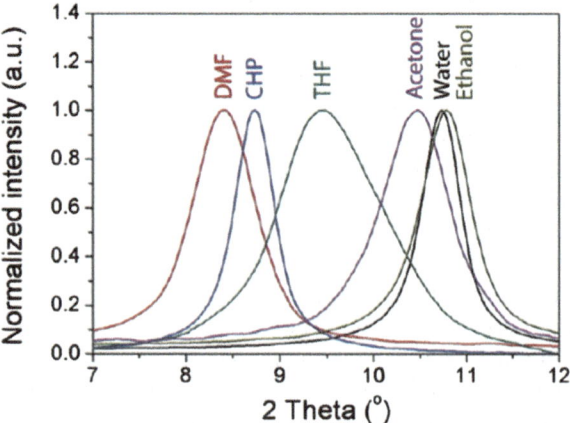

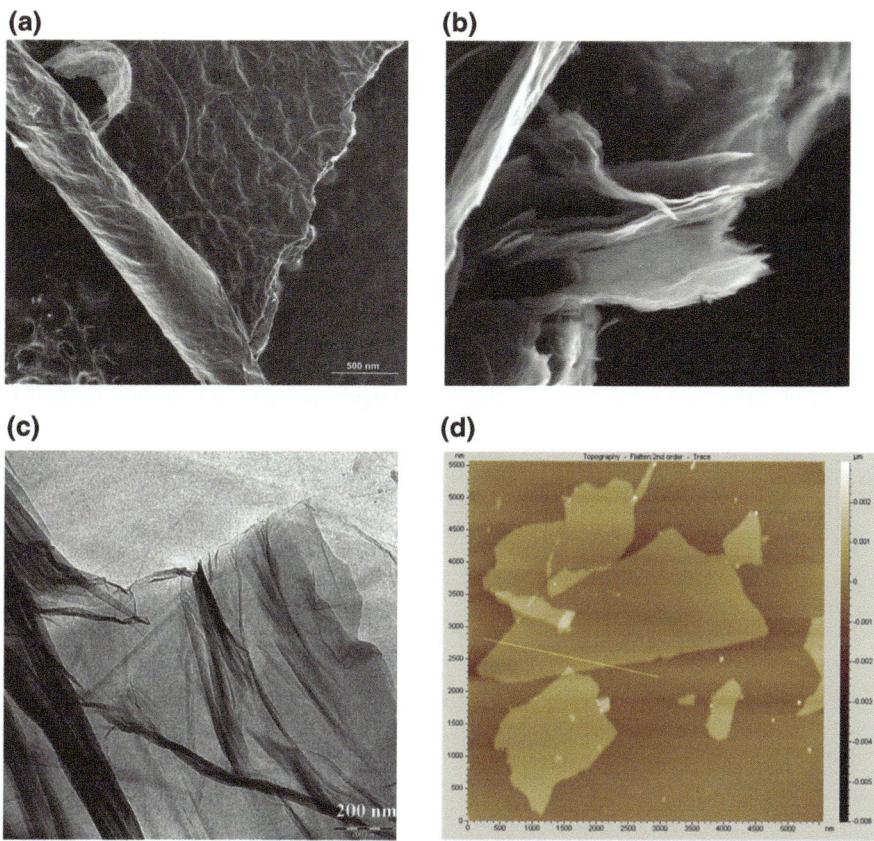

Fig. 2.9 SEM images of **a** layered structure of GO with rounded layer and **b** detail of packed GO structure. Representative TEM images of GO. **c** Reproduced from Ref. [2] with permission from the Royal Society of Chemistry. **d** AFM image of GO. Reprinted with permission from [50]. Copyright (2014) American Chemical Society

XRD. X-ray diffraction (XRD) pattern of pristine graphene oxide shows only a broad peak around 11° [1, 7, 20, 24, 27, 31]. This peak is associated with the interlayer distance ∼1 nm due to the presence of functional groups onto GO. This distance varies depending on the solvent in which GO is dispersed. Jalili et al. [24] reported an interlayer minimum distance of 0.82 nm for ethanol and 1.17 nm in the case GO is dispersed in DMF, as shown in Fig. 2.8.

SEM/TEM. Figure 2.9a, b show SEM images of oven dry GO. The morphology of GO appears as a tightly packed layers with a corrugate surface that sometimes is wrinkled [35].

The TEM image is useful to identify a single layer of graphene oxide, as shown in Fig. 2.9c. This technique shows highly electron transparent corrugated or wrinkled

Fig. 2.10 a DSC profiles for
GO under air and N_2
atmosphere at $10\,K\,min^{-1}$.
Reprinted with permission
from [48]. Copyright (2008)
American Chemical Society.
b TGA profiles for graphite,
graphene and graphene oxide
in dry air. Reprinted from
[37], Copyright (2013), with
permission from Elsevier

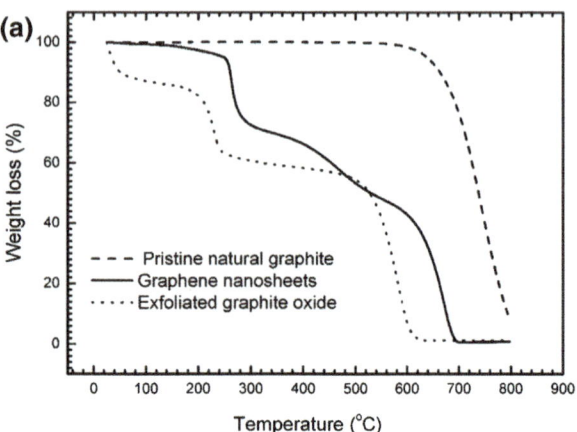

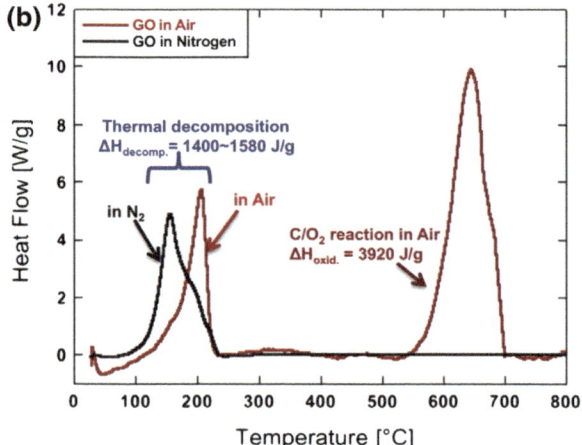

structure of GO layer. Typical examples of GO single and multilayers are reported
by Aunkor et al. [2] and Wang et al. [48].

AFM. Atomic force microscopy (AFM) is able to characterize the lateral size and
the thickness of GO layer. Typically, the height profile reveals a thickness of 1–
1.2 nm while the lateral size can range in the order of tens to hundreds micrometers,
depending on the synthesis and the post-synthesis treatment, e.g. sonication. A representative AFM image is shown in Fig. 2.9d, where it can be noted that the lateral size
of GO single layer ranging from 500 nm to 50 μm. It can be observed an overlapping
of multiple layers with a height of 1 nm for each step [50].

Thermogravimetric Analysis (TGA). The thermal decomposition of GO is shown
in Fig. 2.10a and reported by Wang et al. [48]. Despite the oxygen concentration, GO
is generally unstable to the temperature with starting around 60 °C. The material are

lost about the 60% of its weight up to 300 °C and lose almost all its mass reaching 900–1000 °C. A fast heating process causes a detonation.

Differential Scanning Calorimetry (DSC). A DSC profiles for GO in air and N_2 was reported by Qiu et al. [37] in Fig. 2.10b. Few information can get from DSC of GO up to 800 °C. The main signal appears over 200 °C with exothermal effect which corresponds to an explosion under nitrogen atmosphere. Authors measured an enthalpy of 1600 J g^{-1} for thermal decomposition under nitrogen and 3920 J g^{-1} for the second exothermic peak under air pressure.

2.2.1 Models and Modelling

Many efforts have been devoted to the structural description of graphene oxide (Fig. 2.11) and several models were proposed [13, 53]. Since the first description, Hoffman and Holst proposed to locate on the basal plane of graphite, with sp^2 hybridization, net molecular form C_2O, made by epoxy groups. Later, Ruess reformulated this model and, considering the presence of hydrogen in GO species, introduced hydroxyl moieties in the basal plane of graphite. With this modifications, this model, which acquired a sp^3 character, is considered formed by a repeat unit where 1/4 of cyclohexanes with epoxide groups localized in the 1, 3 positions and hydroxilated in 4 position. In the Scholz and Boehm model the epoxide and ether groups have been substituted by quinoidal species in a corrugated backbone. In these

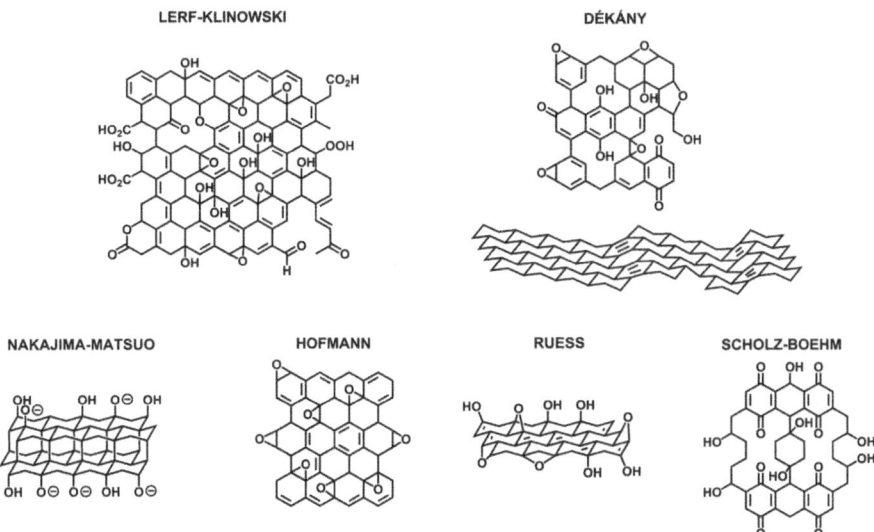

Fig. 2.11 Example of proposed graphene oxide structure. Reproduced from Ref. [13] with permission from The Royal Society of Chemistry

models, however, a misinterpretation is present: the graphene oxide is treated as a materials built up by repetitive units. Lerf and Klinowski abandoned the assumption of the GO periodicity and expressed the model in which the structure is composed by a random distribution of aromatic and wrinkled regions. After all, it is difficult to establish an unique structure for GO because of strictly connected with the synthetic methods.

Besides the interpretation of experimental data to elucidate the GO structure, theoretical studies were carried out approaching towards the fully understand of issue. The first key problem, in a computational study of GO, is related to size of graphene basal plane that usually is too large to be calculated with precision. Several algorithms and methods were referred to the GO structure theoretically. Samarakoon and Wang [38] identified a twist-boat conformation by DFT calculations for a fully-oxidized GO layer with randomly decorated hydroxyl and epoxide groups. Paci et al. [32] explored the formation fo GO structure by means of Monte Carlo method. Here, epoxide and hydroxyl functional groups dominate and are randomly distributed on both side of the graphene plane. They found a set of hydroxyl-hydroxyl and hydroxyl-epoxide hydrogen-bonding interactions and, occasionally, defects made by small holes. In addition, carbonyl and alcohol groups are even molecule of water were observed. Fonseca et al. [14] examined three classical force fields methods. They used Reactive Empirical Bond Order for carbon, third generation of the Charge Optimized Many Body (COMB3), hydrogen and oxygen (REBO-CHO) and Chemistry at HARvard Macromolecular Mechanics (CHARMM) force field to study the properties of GO and to simulate GO structures. Their conclusion is that the COMB3 was considered as the best prediction for almost all properties to investigate physical and chemical properties of different GO structures. Recently, Sk et al. [42] looked into the oxidative process of graphene at edges using DFT. Their results showed that the oxidation is more favourable along the edges comparing with the central part of the graphene basal plane. Froning et al. [15] employed a density functional theory calculations to confirm experimental data for the oxidation process of graphene. They studied the progressive oxidation, using UV/ozone, of a graphene surface by AFM and then compared with DFT and MD simulation to explain that the oxidation starts is susceptible of graphene edges.

References

1. Aboutalebi SH, Gudarzi MM, Zheng QB, Kim JK (2011) Spontaneous formation of liquid crystals in ultralarge graphene oxide dispersions. Adv Funct Mater 21(15):2978–2988
2. Aunkor MTH, Mahbubul IM, Saidur R, Metselaar HSC (2015) Deoxygenation of graphene oxide using household baking soda as a reducing agent: a green approach. RSC Adv 5(86):70461–470472
3. Bagri A, Mattevi C, Acik M, Chabal YJ, Chhowalla M, Shenoy VB (2010) Structural evolution during the reduction of chemically derived graphene oxide. Nat Chem 2(7):581–587
4. Bosch-Navarro C, Busolo F, Coronado E, Duan Y, Martí-Gastaldo C, Prima-Garcia H (2013) Influence of the covalent grafting of organic radicals to graphene on its magnetoresistance. J

Mater Chem C 1(30):4590–9

5. Brodie BC (1859) On the atomic weight of graphite. Phil Trans R Soc Lond 149:249–259

6. Chen D, Feng H, Li J (2012) Graphene oxide: preparation, functionalization, and electrochemical applications. Chem Rev 112(11):6027–6053

7. Chen J, Li Y, Huang L, Li C, Shi G (2015) High-yield preparation of graphene oxide from small graphite flakes via an improved Hummers method with a simple purification process. Carbon 81:826–834

8. Chen ZL, Kam FY, Goh RGS, Song J, Lim GK, Chua LL (2013) Influence of graphite source on chemical oxidative reactivity. Chem Mater 25(15):2944–2949

9. Chua CK, Pumera M (2014) Chemical reduction of graphene oxide: a synthetic chemistry viewpoint. Chem Soc Rev 43(1):291–312

10. Chung MG, Kim DH, Lee HM, Kim T, Choi JH, Dk S, Yoo JB, Hong SH, Kang TJ, Kim YH (2012) Highly sensitive NO2 gas sensor based on ozone treated graphene. Sens Actuators B Chem 166–167:172–176

11. Dimiev AM, Tour JM (2014) Mechanism of graphene oxide formation. ACS Nano 8(3):3060–3068

12. Dreyer DR, Park S, Bielawski CW, Ruoff RS (2010) The chemistry of graphene oxide. Chem Soc Rev 39(1):228–240

13. Dreyer DR, Todd AD, Bielawski CW (2014) Harnessing the chemistry of graphene oxide. Chem Soc Rev 43(15):5288–5301

14. Fonseca AF, Liang T, Zhang D, Choudhary K, Sinnott SB (2016) Probing the accuracy of reactive and non-reactive force fields to describe physical and chemical properties of graphene-oxide. Comput Mater Sci 114:236–243

15. Froning JP, Lazar P, Pykal M, Li Q, Dong M, Zboril R, Otyepka M (2017) Direct mapping of chemical oxidation of individual graphene sheets through dynamic force measurements at the nanoscale. Nanoscale 9(1):119–127

16. Gao W (2015) Graphene oxide: reduction recipes, spectroscopy, and applications. Springer: New York

17. Georgakilas V (2014) Functionalization of graphene. Wiley, New York

18. Gilje S, Han S, Wang M, Wang KL, Kaner RB (2007) A chemical route to graphene for device applications. Nano Lett 7(11):3394–3398

19. Guerrero-Contreras J, Caballero-Briones F (2015) Graphene oxide powders with different oxidation degree, prepared by synthesis variations of the Hummers method. Mater Chem Phys 153:209–220

20. Hong Y, Wang Z, Jin X (2013) Sulfuric acid intercalated graphite oxide for graphene preparation. Sci Rep 3:132

21. Hossain MZ, Johns JE, Bevan KH, Karmel HJ, Liang YT, Yoshimoto S, Mukai K, Koitaya T, Yoshinobu J, Kawai M, Lear AM, Kesmodel LL, Tait SL, Hersam MC (2012) Chemically homogeneous and thermally reversible oxidation of epitaxial graphene. Nat Chem 4(4):305–309

22. Huh S, Park J, Kim YS, Kim KS, Hong BH, Nam JM (2011) UV/ozone-oxidized large-scale graphene platform with large chemical enhancement in surface-enhanced Raman scattering. ACS Nano 5(12):9799–9806

23. Hummers WS, Offeman RE (1958) Preparation of graphitic oxide. J Am Chem Soc 80(6):1339–1339

24. Jalili R, Aboutalebi SH, Esrafilzadeh D, Konstantinov K, Moulton SE, Razal JM, Wallace GG (2013) Organic solvent-based graphene oxide liquid crystals: a facile route toward the next generation of self-assembled layer-by-layer multifunctional 3D architectures. ACS Nano 7(5):3981–3990

25. Jung I, Field DA, Clark NJ, Zhu Y, Yang D, Piner RD, Stankovich S, Dikin DA, Geisler H, Ventrice Jr CA, Ruoff RS (2009) Reduction kinetics of graphene oxide determined by electrical transport measurements and temperature programmed desorption. J Phys Chem C 113(43):18480–18486

26. Kovtyukhova NI, Ollivier PJ, Martin BR, Mallouk TE, Chizhik SA, Buzaneva EV, Gorchinskiy AD (1999) Layer-by-layer assembly of ultrathin composite films from micron-sized graphite oxide sheets and polycations. Chem Mater 11(3):771–778
27. Krishnamoorthy K, Veerapandian M, Yun K, Kim SJ (2013) The chemical and structural analysis of graphene oxide with different degrees of oxidation. Carbon 53:38–49
28. Kudin KN, Ozbas B, Schniepp HC, Prud'homme RK, Aksay IA, Car R (2008) Raman spectra of graphite oxide and functionalized graphene sheets. Nano Lett 8(1):36–41
29. Lai Q, Zhu S, Luo X, Zou M, Huang S (2012) Ultraviolet-visible spectroscopy of graphene oxides. AIP Adv 2(3):032146
30. Ma HL, Zhang HB, Hu QH, Li WJ, Jiang ZG, Yu ZZ, Dasari A (2012) Functionalization and reduction of graphene oxide with p-phenylene diamine for electrically conductive and thermally stable polystyrene composites. ACS Appl Mater Interfaces 4(4):1948–1953
31. Marcano DC, Kosynkin DV, Berlin JM, Sinitskii A, Sun Z, Slesarev A, Alemany LB, Lu W, Tour JM (2010) Improved synthesis of graphene oxide. ACS Nano 4(8):4806–4814
32. Paci JT, Belytschko T, Schatz GC (2007) Computational studies of the structure, behavior upon heating, and mechanical properties of graphite oxide. J Phys Chem C 111(49):18099–18111
33. Pendolino F, Capurso G, Maddalena A, Lo Russo S (2014a) The structural change of graphene oxide in a methanol dispersion. RSC Adv 4(62):32,914
34. Pendolino F, Parisini E, Lo Russo S (2014b) Time-dependent structure and solubilization kinetics of graphene oxide in methanol and water dispersions. J Phys Chem C 118(48):28162–28169
35. Pendolino F, Armata N, Masullo T, Cuttitta A (2015) Temperature influence on the synthesis of pristine graphene oxide and graphite oxide. Mater Chem Phys 164:71–77
36. Peng L, Xu Z, Liu Z, Wei Y, Sun H, Li Z, Zhao X, Gao C (2015) An iron-based green approach to 1-h production of single-layer graphene oxide. Nat Comms 6:5716
37. Qiu Y, Collin F, Hurt RH, Külaots I (2016) Thermochemistry and kinetics of graphite oxide exothermic decomposition for safety in large-scale storage and processing. Carbon 96:20–28
38. Samarakoon DK, Wang XQ (2011) Twist-boat conformation in graphene oxides. Nanoscale 3(1):192–195
39. Schniepp HC, Li JL, McAllister MJ, Sai H, Herrera-Alonso M, Adamson DH, Prud'homme RK, Car R, Saville DA, Aksay IA (2006) Functionalized single graphene sheets derived from splitting graphite oxide. J Phys Chem 110(17):8535–8539
40. Scholz W, Boehm HP (1969) Untersuchungen am Graphitoxid. VI. Betrachtungen zur Struktur des Graphitoxids. Z Anorg Allg Chem 369(3–6):327–340
41. Shin HJ, Kim KK, Benayad A, Yoon SM, Park HK, Jung IS, Jin MH, Jeong HK, Kim JM, Choi JY, Lee YH (2009) Efficient reduction of graphite oxide by sodium borohydride and its effect on electrical conductance. Adv Funct Mater 19(12):1987–1992
42. Sk MA, Huang L, Chen P, Lim KH (2016) Controlling armchair and zigzag edges in oxidative cutting of graphene. J Mater Chem C 4(27):6539–6545
43. Stankovich S, Da D, Piner RD, Kohlhaas KA, Kleinhammes A, Jia Y, Wu Y, Nguyen ST, Ruoff RS (2007) Synthesis of graphene-based nanosheets via chemical reduction of exfoliated graphite oxide. Carbon 45(7):1558–1565
44. Starodub E, Bartelt NC, McCarty KF (2010) Oxidation of graphene on metals. J Phys Chem C 114(11):5134–5140
45. Staudenmaier L (1898) Verfahren zur Darstellung der Graphitsäure. Ber Dtsch Chem Ges 31(2):1481–1487
46. Sun L, Fugetsu B (2013) Mass production of graphene oxide from expanded graphite. Mater Lett 109:207–210
47. Vinogradov NA, Schulte K, Ng ML, Mikkelsen A, Lundgren E, Mårtensson N, Preobrajenski AB (2011) Impact of atomic oxygen on the structure of graphene formed on Ir(111) and Pt(111). J Phys Chem C 115(19):9568–9577
48. Wang G, Yang J, Park J, Gou X, Wang B, Liu H, Yao J (2008) Facile synthesis and characterization of graphene nanosheets. J Phys Chem C 112(22):8192–8195

49. Yamamoto M, Einstein TL, Fuhrer MS, Cullen WG (2012) Charge inhomogeneity determines oxidative reactivity of graphene on substrates. ACS Nano

50. Yang H, Li H, Zhai J, Sun L, Yu H (2014) Simple synthesis of graphene oxide using ultrasonic cleaner from expanded graphite. Ind Eng Chem Res 53(46):17878–17883

51. You S, Luzan SM, Szabó T, Talyzin AV (2013) Effect of synthesis method on solvation and exfoliation of graphite oxide. Carbon 52:171–180

52. Zhang Y, Ma HL, Zhang Q, Peng J, Li J, Zhai M, Yu ZZ (2012) Facile synthesis of well-dispersed graphene by γ-ray induced reduction of graphene oxide. J Mater Chem 22(26):13064

53. Zhao J, Liu L, Li F (2015) Graphene oxide: physics and applications. Springer Briefs in Physics. Springer, Berlin

Chapter 3
Regulation and Environmental Aspects of Graphene Oxide

Abstract Applications of smart nanomaterials are expected to impact many aspects of life and promoting a substantial and advanced change in technologies. However, recently the toxicity of these very promising materials and their environmental impact has been questioned. Graphene oxide is not yet regulated by a specific normative, thus the parent definition for nanomaterials is adapted. This chapter presents the normative and regulations on the nanomaterials with particular attention for the EU area. Because of a legislative framework is not yet presented or incomplete, guidelines and definitions support normative in view of auxiliary results from scientific progress on smart nanomaterials. In addition, a provisorily ISO technical document is approaching to define and standardize the producers to work with a nano-object. As graphene oxide is a novel nanomaterial, environmental aspects, connected to its employment, are now limited and further scientific understanding is needed to establish the eventual health and ecosystem risks.

Keywords Normative · Nanomaterials · Toxicity · Environment · Disposal

3.1 Regulation on Nanomaterials

In recent years, manufactures made or containing nanomaterials are available on the market in almost any industrial sectors. Nevertheless, a legislative framework that covers nanomaterials is lacking and more specific research is needed to address regulatory questions accounting for the definition of nanomaterial, safe handling methods and risk assessment [39]. Several countries as Australia [1] and Canada [23], have published a working definition or presented guidelines as USA (US-FDA) [44] on nanomaterials. Since 2011, the EU have acted with a "Recommendation on the definition of a nanomaterial" [14] that is based on references reports by European Commission Joint Research Centre in 2010 [38] and scientific opinion by the SCENIHR in 2009 [9]. The European Commission reviewed the definition on 2012 and is expected to conclude the revision by 2016. The purpose of the EU recommendation is to enable conformity across different legislative areas and to

© The Author(s) 2017   23
F. Pendolino and N. Armata, *Graphene Oxide in Environmental Remediation Process*, SpringerBriefs in Applied Sciences and Technology, DOI 10.1007/978-3-319-60429-9_3

treat a nanomaterial as such over several sectors. According to the EU definition of *Nanomaterials*:

> Nanomaterial means a natural, incidental or manufactured material containing particles, in an unbound state or as an aggregate or as an agglomerate and where, for 50% or more of the particles in the number size distribution, one or more external dimensions is in the size range 1–100 nm.
>
> In specific cases and where warranted by concerns for the environment, health, safety or competitiveness the number size distribution threshold of 50% may be replaced by a threshold between 1 and 50%.
>
> By derogation from the above, fullerenes, graphene flakes and single wall carbon nanotubes with one or more external dimensions below 1 nm should be considered as nanomaterials.

This definition appears a rather broad description for nanomaterials and can include all materials, natural or synthetic, with a size below 100 nm. The main purpose of the EU definition is to provide a clear and unambiguous criteria to identify a nanomaterial and to differentiate it from their corresponding bulk. Nevertheless, some fragmented aspects are found in the definition. In fact, the definition covers nanomaterials from natural sources, such as ashes, and is lacking in about their origin. To overcome this, it is often taken into account the *size distribution* or the *specific surface area by volume* (VSSA). There are no scientific proofs that justify a threshold value in which the physical chemical properties change at 100 nm, thus a stochastic size distribution results in a more appropriate parameter to describe a nanomaterial. Therefore, in this recommendation when the size distribution is over 50% of nanomaterial/nanoparticles with a size 1–100 nm the whole material is defined as nanomaterial. Alternatively, the use of the *volume specific surface area*, as a definition parameter, includes materials with internal structure. By definition, limit $60 \, m^2/cm^3$ is chosen as threshold value based on a volume specific surface area for a perfect sphere of 100 nm. VSSA greater of $60 \, m^2/cm^3$ indicates a nanomaterials with a size equal or smaller of 100 nm. However, for dry powder the BET method is already established, but for or liquid no standard techniques are employed. A description of techniques for analyzing nanomaterials are listed by JRC [28], ECHA [13], OECD [37] and European Food Safety Agency [12].

A further issue is associated with the *lifetime* of a nanomaterial [6]. In fact, particles may be aggregate or disaggregate switching from nanomaterial range to bulk or viceversa. This aspect is missing in the EU definition of nanomaterial and must be taken into account. For instance, nanomaterial may be included and merged into a scaffold structure losing the identity of nanomaterial because of the particles are not anymore in the range of 1–100 nm

In general, the assumption that smaller means more reactive and thus harmful is not accurate. The *nanomaterials are not intrinsically hazardous* and must be treated as a normal chemical/substance that may be toxic or not. The European Chemical Agency (ECHA) established in 2012 a ECHA-NMWG working group regarding nanomaterials "to discuss scientific and technical questions relevant to REACH[1] and CLP[2] processes and to give recommendations on strategic issues". The ECHA-NMWG's aim is to discuss with industries and allows the REACH requirements. Nevertheless, nanomaterials are not explicitly mentioned within REACH. Today, several pieces of EU legislation are referred to nanomaterials, such as the regulation of the use of biocidal products (528/2012/EU) [18]. Here, in the preamble point 66, the European Parliament states:

> There is scientific uncertainty about the safety of nanomaterials for human health, animal health and the environment. In order to ensure a high level of consumer protection, free movement of goods and legal certainty for manufacturers, it is necessary to develop an uniform definition for nanomaterials, if possible based on the work of appropriate international forums and to specify that the approval of an active substance does not include the nanomaterial form unless explicitly mentioned. The Commission should regularly review the provisions on nano materials in the light of scientific progress.

A related issue concerns the specific risk of dispersing/dissolving of nanomaterial in an environmental or physiological medium and a secondary change to a non-nanomaterial. Therefore, it becomes clear that the European Commission required further scientific investigations before the "nano" is related to health risks.

3.1.1 Regulation on Graphene Oxide

The current legislation does not have a specific normative on graphene oxide. Thus, the parent normative for nanomaterials can be adapted. Scientifically, graphene oxide is defined as a nanomaterial. However, further studies are needed to solve a number of issues, such as the GO composition and the lateral size. According to EU definition (Sect. 3.1), a nanomaterial has a size in the range 1–100 nm, but the size of GO is variable and is controlled by the grain size of carbon source and by the synthetic method. The lateral size of graphene oxide varies from $0.1\,\mu$m up to $300\,\mu$m and even over this value. This means that for the lower value of lateral size ($\sim0.1\,\mu$m) graphene oxide can be classified as a nanomaterial, while above the upper value, graphene oxide falls within micro material or bulk. Despite the variety of lateral sizes, "macro" graphene oxide exhibits a very similar behaviour comparable with GO in the nano range. For some materials, such as graphene oxide, the upper limit of 100 nm is not scientifically correlated with "nano". Alternatively, as suggested by

---

[1]Registration, Evaluation, Authorisation and Restriction of Chemicals (REACH) regulation.
[2]Classification, Labelling and Packaging (CPL) regulation.

the EU, the size distribution or the volume specific surface area (VSSA) can be used to identify a nanomaterial. However, both methods are some critical issues. The size distribution of the graphene oxide is strongly affected by the post-synthesis treatment, such as sonication, which alters the lateral size from micrometer to nanometer range. This means that GO falls down to the nanomaterial definition, without altering the general physical chemical properties.

By concerning the volume specific surface area, the BET analysis ranges from $5\,m^2/cm^3$ to about $1900\,m^2/cm^3$, with average of $500\,m^2/cm^3$. This broad range depends on the dry procedure to get solid GO. For instance, by drying in an oven, a layered arrangement is produced, while the freezing cast method gets a sponge structure. Because of the VSSA threshold value for nanomaterial is defined as $60\,m^2/cm^3$, the same GO can be a nano or macro material depending on the dry procedure [10, 26, 41]. For this reason, the EU made an exception for some carbon allotropes and by derogation, materials with at least "one dimension below 1 nm should be considered as a nanomaterial". Graphene oxide possesses a height of about 1 nm, hence is accounted as a nanomaterial.

Up to date, graphene oxide has not a CAS number or REACH registration. In the majority of the scientific literature, graphene oxide is lab synthesized and only few companies sell it (ACS Materials, Sigma-Aldrich, Avanzare, Graphanea, Graphene Supermarket, etc.). Because of graphene oxide is still a new product, details about the risk, symptoms/effects and safety information are poor. Safety Data Sheet (SDS) for graphene oxide refers to starting materials, graphite and water, because of lacking information on pristine graphene oxide. The comparison of graphene oxide with graphite is lacking of information and brings to no correct conclusion on material. Graphite is a fine powder and volatile, while dry graphene oxide forms a layered substance. This physical chemical property limits the inhalation of GO. Ingestion or toxicity for graphite is greater of 2000 mg/kg (the LD50 (oral, rat)). A threshold LD50 is not yet found for GO and contrasting results in scientific literature are reported [3–5, 19–22, 31]. In general, toxicity of graphene oxide appears concentration-dependent with values that vary from tenths to hundreds of GO μg/mL [8, 11, 45]. Pristine graphene or reduced graphene oxide induces a more toxic effect than graphene oxide and higher biocompatibility were shown in GO possessing low oxygen content [32]. Nevertheless, there are too little research and no standard criteria with enough detailed information that is impossible to come to an explicit conclusion about the GO toxicity. Further studies are needed on the toxicity effect on humans or animals. Exposure of the eyes or skin to graphene oxide does not show the known effect.

The only safety note, concerning GO, consists in the reactivity of dry solid graphene oxide with a temperature over 60° or close to the flame. At this condition, GO starts to decompose and, in the presence of potassium cations, self-ignites and detonates [27].

Because of the novelty of graphene oxide material, any relevant legislation is nowadays explicit concern on this derivative of graphene. The International Organization for Standardization (ISO) is working on technical document (ISO/TS 80004) concerns the *nanotechnologies* following the advancement across the scientific communities to define a "nano" and its broad meaning. The ISO/TS 80004 is including vocabulary definition, healthcare, and characterization techniques. Special parts are under preparation and, in particular, the *Part 13: Graphene and other two-dimensional materials*, in which will be discussed the standard by ISO technical committee.

3.2 Environmental Aspects

As mentioned above, nanomaterials are now included in a widely available products to enhance the properties due to their nano scale structure. Graphene oxide is among these materials and can be treated by the normative as a nanomaterials. Associating with their uses, nanomaterials are found in the waste and the possible risks to environmental health and safety. Various organizations have paid attention on the lack of understanding and reported on the human health and environmental risks of nanomaterial should be evaluated. Nevertheless, a policy framework is not established and legislation for disposal of nanomaterial si under evaluation by countries following the scientific advancements. The real impact of nanomaterials on environment and health risk is under studies and is in its embryonal stage. Since 2009, the EU reviewed a number of aspects of the waste legislation due to the presence of nanomaterials. The conclusion of the Commission demonstrated the difficulty of separating or eliminating nanomaterials form waste [24]. The protection of the environment was introduced by the EU within Environmental Liability Directive (Directive 2004/35/EC) and was implemented across the EU in 2010 [16]. This legal reference has the purpose of establishing a framework of environmental liability and prevent environmental damage.

3.2.1 Environmental GO Toxicity

Toxicity of graphene oxide is a crucial issue when we consider its use in environmental remediation. Several works are evaluated for technological applications, nevertheless, a full comprehension is not risen and, sometimes, contrasting results are reported. It was observed that the morphology, the lateral dimension and the oxygen content play a key role. Additional details can be found in reviews [11, 43]. The antibacterial and cytotoxicity are the main issues that matter in environmental remediation and the latest discoveries are detailed below.

Morphology of GO flakes associated with the bacterial toxicity, was reported by Akhavan et al. [3]. They argued that the bacterial inactivation was caused by the contact of the bacteria cell membrane with the sharp edges of the GO. For this reason, the *Escherichia coli* bacteria, having an outer membrane, reveal a greater resistance respect to Staphylococcus aureus, having no outer membrane. In addition, the reduction of GO produces a more toxic material probably because of a sharper edges, as claimed by authors. Liu et al. [30] understood the influence of the GO

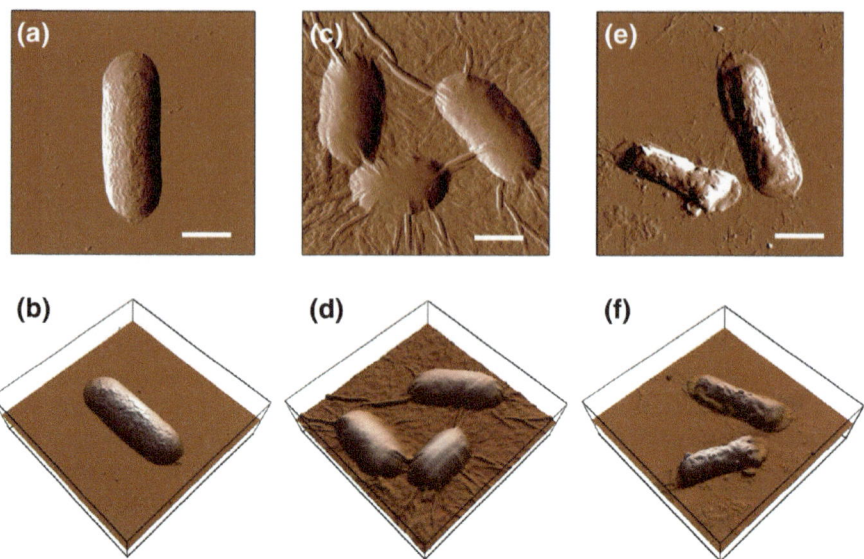

Fig. 3.1 AFM and 3D images of *Escherichia coli* bacteria after incubation with GO. **a, b** incubation with deionized water for 2 h; **c, d** incubation with the 40 μg/mL GO-0 for 2 h; **e, f** incubation with the 40 μg/mL GO-240 for 2 h. The mark is 1 μm. Reprinted with permission from [30]. Copyright (2012) American Chemical Society

lateral size towards a model bacterium, *Escherichia coli* (E. coli). The antibacterial activity of GO is lateral-size dependent and increases with the size of GO flake. As seen in Fig. 3.1, the E. coli was incubated with deionized water without showing any change (a, b), while the cell surface roughness decreases using large size GO flake (c, d). When small size GO covers the cell surface (e, f) the surface roughness increments. Authors argue that when the surface of GO is enough large, the bacterial membrane is covered and nutrients are blocked, hence the membrane may become biologically inactive.

Liu et al. [29] compared graphene based material: graphene oxide, reduced graphene oxide (rGO), graphite oxide (GtO) and graphite (Gt). Authors's results showed the highest antibacterial activities of GO towards *Escherichia coli* bacteria, followed by rGO, Gt, and GtO. They assert that the interaction with bacterial cells is affected by the density of functional groups and smaller in size of graphene oxide. Recently, the oxygen content in graphene oxide was investigated by Masullo et al. [32]. They compared two types of GO possessing different degree of oxidation, i.e. 20 and 50%. The in vitro hemolytic activity reveals a more biocompatible material for what concerning the less oxidized graphene oxide species (GO 20%). Comparable results took place at graphite oxide (oxygen content at 20%) and graphite (particle size 45 μm). On opposite, a negative hemolysis assay come up with graphene oxide with higher oxygen content (GO 50%). The acute toxicity, short-term and high load, of GO in wastewater microbial communities was investigated by Ahmed et al. [2].

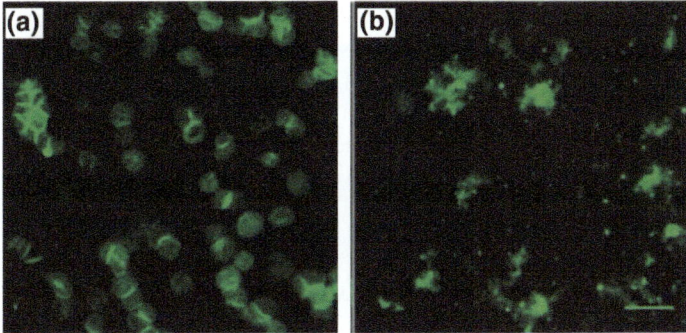

Fig. 3.2 Typically confocal images of **a** erythrocytes and, **b** after injection with GO. Reprinted from Publication [8]. Copyright (2016), with permission from Elsevier

They showed a concentration-dependence of GO on the toxicity of microbial communities, in particular in the range 50–300 mg/L. Authors conclude that the presence of GO remarkably affects the bacterial metabolic activity and the presence of GO in the activated sludge led to reduced BOD_5. This results confirms that GO hinders the metabolic activity to end up with the cell death. The in vitro toxicity study on human lung cells (normal as well as cancer) was conducted by Mittal et al. [33] very recently. In this study, the cytotoxicity of graphene oxide and two derivates, thermal reduced GO and chemical reduced GO, are examined toward bronchial epithelial cells (BEAS-2B) and alveolar epithelial cells (A549). Authors demonstrated that the lateral size and functional groups are likely responsible for the behaviour of graphene oxide and derivates in human lung cells. The major effect was found for thermal reduced GO compared to chemical reduced GO and pristine GO. Thus, the toxicity intensity was proportional with the reduced lateral size and increased functional groups. The biocompatibility of the red blood cells with GO was considered by the study of De Spirito et al. [8]. Their conclusion infers that the GO flakes are able to damage the erythrocyte plasma membrane and the haemolytic activity is inversely proportional with lateral size of GO. In Fig. 3.2 is shown confocal images where the RBCs rounded shape (a) is lost by adding GO (b). Nevertheless, other researchers have reported different conclusions for the GO toxicity. For instance, Ruiz et al. [40] reported the antibacterial activity and its biocompatibility with mammalian cells in the graphene oxide. In this study, the inclusion of GO caused a fast bacteria grown than cultures without GO. Authors so come to an end that graphene oxide does not possess antibacterial, bacteriostatic, and cytotoxic properties in both bacteria and mammalian cells. Recently, quantitative structure-activity relationship (QSAR) models can predict the relationship between toxicity and structure of particles at nanoscale [7]. The ECHA promotes the use of this approach to deliver reliable information on registered molecules. To the best of our knowledge, any articles are reported a study on the use of QSAR model for toxicity of GO or toxicity of target molecule/GO composite in drug delivery systems.

Up to date, we cannot draw a general conclusion of the impact of graphene oxide on environmental toxicity and too little is known about the interaction between GO and bacteria or immune systems. Many of the issues come from the fact that physical chemical properties of GO are not standardized due to the non-stoichiometric chemical structure. Further researches are needed to clarify this important aspect before graphene oxide can be implemented into a commercial application.

3.2.2 Disposal of Nanomaterials

Nanomaterial waste should be considered and categorised as a conventional waste because there is not a specific legislation. As suggested by the Nanotechnology Industrial Association (NIA) [34, 35], wastes that contain nanomaterials can be processed as a waste originated (i) from the production; (ii) during the use; or (iii) from its end-of-life activity, such as recycling. For a general approach, the UKs Control of Substances Hazardous to Health Regulation (COSHH) [25], recommends to safe handling and control of nanomaterials by identifying the risks, prevent/controlling the exposure, monitoring the exposure, carrying out the health surveillance and deciding the precautions to take, planning procedure to deal with emergencies and, finally, inform and train the employees.

The EU acts on treatment of waste with the Directive 2008/98/EC (Waste Framework Directive) [17] and in 2011 published "Coherence of Waste Legislation" [15] to implement new materials, such as nanomaterials, in EU waste policy. On this base, independent organisations published a technical report and guidelines to disposal objects containing nanomaterials. On 2012, the British Standards Institute (BSI) reported on "Disposal of manufacturing process waste containing manufactured nano-objects. Guide" [42]. The guide provides technical practice to process waste containing "unbound manufactured and/or engineered nano-objects, including agglomerates and aggregates of the same" and an appropriate routes to prepare waste to transfer to a disposal company. Here, it is included a decision flow chart to assist users and applying the recommendations.

Moreover, Organisation for Economic Cooperation and Development (OECD) has published a book on "Nanomaterials in Waste Streams" [36]. This book considers the diversity of waste treatments based on the waste composition and deals with recycling, incineration, land filling and wastewater treatment. The OECD concludes that little is known to date concerning the waste treatment of materials contains nano-objects. Even if we possess enough technology to capture or eliminate nanomaterials there are concerns about health risk and release into environment, especially for the agriculture. Further scientific knowledges are needed for a safe handling of nanomaterials.

References

1. National Industrial Chemicals Notification and Assessment Scheme—NICNAS (2010) Adjustments to Nicnas new chemicals processes for industrial nanomaterials. Chemical Gazette C 10
2. Ahmed F, Rodrigues DF (2013) Investigation of acute effects of graphene oxide on wastewater microbial community: a case study. J Hazard Mater 256–257:33–39
3. Akhavan O, Ghaderi E (2010) Toxicity of graphene and graphene oxide nanowalls against bacteria. ACS Nano 4(10):5731–5736
4. Atta NF, Galal A, El-Ads EH (2015) Graphene—a platform for sensor and biosensor applications. In: Toonika R (ed) Biosensors-micro and nanoscale applications, InTech, pp 37–84
5. Bianco A (2013) Graphene: safe or toxic? The two faces of the medal. Angew Chem Int Ed 52(19):4986–4997
6. Bleeker EAJ, de Jong WH, Geertsma RE, Groenewold M, Heugens EHW, Koers-Jacquemijns M, van de Meent D, Popma JR, Rietveld AG, Wijnhoven SWP, Cassee FR, Oomen AG (2013) Considerations on the EU definition of a nanomaterial: science to support policy making. Regul Toxicol Pharmacol 65(1):119–125
7. Burello E, Worth A (2011) Computational nanotoxicology: predicting toxicity of nanoparticles. Nat Nanotechnol 6(3):138–139
8. De Spirito M, Papi M, Maolucci G, Ciasca G (2016) Plasma protein corona reduces the haemolytic activity of the graphene oxide nano and micro flakes. Biophys J 110(3):167a
9. Directorate C: Public Health and Risk Assessment (2010) Scientific Basis for the Definition of the Term "nanomaterial". Technical report. Scientific Committee on Emerging and Newly Identified Health Risks (SCENIHR)
10. Dreyer DR, Park S, Bielawski CW, Ruoff RS (2010) The chemistry of graphene oxide. Chem Soc Rev 39(1):228–240
11. Dudek I, Skoda M, Jarosz A, Szukiewicz D (2015) The molecular influence of graphene and graphene oxide on the immune system under in vitro and in vivo conditions. Arch Immunol Ther Exp 64(3):195–215
12. EFSA J (2011) Guidance on the risk assessment of the application of nanoscience and nanotechnologies in the food and feed chain 9(5):2140
13. European Chemicals Agency ECHA (2012) Appendix R7–1 Recommendations for nanomaterials applicable to Chapter R7a Endpoint specific guidance. Technical report. European Chemical Agency (ECHA)
14. European Commission (DG ENV) (2011a) Commission recommendation of 18 October 2011 on the definition of nanomaterial. OJ L275:38–40
15. Commission European, (DG ENV) (2011b) Study on coherence of waste legislation. Technical report. Bio Intelligence Service
16. European Parliament and the Council (2004) Directive 2004/35/CE of the European Parliament and of the council of 21 April 2004 on environmental liability with regard to the prevention and remedying of environmental damage OJ:56–75
17. OJ L (2008) Directive 2008/98/EC of the European Parliament and of the council of 19 November 2008 on waste and repealing certain directives 312:3–30
18. OJ L (2012) Regulation (EU) No 528/2012 of the European Parliament and of the council of 22 May 2012 concerning the making available on the market and use of Biocidal products 167:1
19. Feng L, Wu L, Qu X (2012) New horizons for diagnostics and therapeutic applications of graphene and graphene oxide. Adv Mater 25(2):168–186
20. Gao J, Bao F, Feng L, Shen K, Zhu Q, Wang D, Chen T, Ma R, Yan C (2011) Functionalized graphene oxide modified polysebacic anhydride as drug carrier for levofloxacin controlled release. RSC Adv 1(9):1737
21. Georgakilas V, Tiwari JN, Kemp KC, Perman JA, Bourlinos AB, Kim KS, Zboril R (2016) Noncovalent functionalization of graphene and graphene oxide for energy materials, biosensing, catalytic, and biomedical applications. Chem Rev 116(9):5464–5519

22. Gurunathan S, Woong Han J, Kim E, Kwon DN, Park JK, Kim JH (2014) Enhanced green fluorescent protein-mediated synthesis of biocompatible graphene. J Nanobiotechnol 12(1):41
23. Health Canada (2017) policy statement on health Canada's working definition for nanomaterial. http://www.hc-sc.gc.ca/sr-sr/pubs/nano/pol-eng.php. Accessed 30 Apr 2017
24. Hull M, Bowman D (2014) Nanotechnology environmental health and safety: risks, regulation, and management. Elsevier, Oxford
25. IOM SAFENANO (2017) Current guidance for safe handling and control of nanomaterials. www.safenano.org/knowledgebase/guidance/safehandling/. Accessed 30 Apr 2017
26. Kaniyoor A, Baby TT, Ramaprabhu S (2010) Graphene synthesis via hydrogen induced low temperature exfoliation of graphite oxide. J Mater Chem 20(39):8467–8469
27. Krishnan D, Kim F, Luo J, Cruz-Silva R, Cote LJ, Jang HD, Huang J (2012) Energetic graphene oxide: challenges and opportunities. Nano Today 7(2):137–152
28. Linsinger T, Roebben G, Gilliland D, Calzolai L, Rossi F, Gibson N, Klein C (2012) Requirements on measurements for the implementation of the European Commission definition of the term 'nanomaterial'. Technical report. Joint Research Centre
29. Liu S, Zeng TH, Hofmann M, Burcombe E, Wei J, Jiang R, Kong J, Chen Y (2011) Antibacterial activity of graphite, graphite oxide, graphene oxide, and reduced graphene oxide: membrane and oxidative stress. ACS Nano 5(9):6971–6980
30. Liu S, Hu M, Zeng TH, Wu R, Jiang R, Wei J, Wang L, Kong J, Chen Y (2012) Lateral dimension-dependent antibacterial activity of graphene oxide sheets. Langmuir 28(33):12364–12372
31. Liu X, Chen KL (2015) Interactions of graphene oxide with model cell membranes: probing nanoparticle attachment and lipid bilayer disruption. Langmuir 31(44):12076–12086
32. Masullo T, Armata N, Pendolino F, Colombo P, Celso FL, Mazzola S, Cuttitta A (2015) Low-cost synthesis of smart biocompatible graphene oxide reduced species by means of GFP. Appl Biochem Biotechnol 178(3):462–473
33. Mittal S, Kumar V, Dhiman N, Chauhan LKS, Pasricha R, Pandey AK (2016) Physico-chemical properties based differential toxicity of graphene oxide/reduced graphene oxide in human lung cells mediated through oxidative stress. Sci Rep 6(39):548
34. Nanotechnology Industries Association (NIA) (2013) Closing the gap: the impact of nanotechnologies on the global divide. NIA Report
35. Nanotechnology Industries Association (NIA) (2017) Recycling and Waste. www.nanotechia.org/sectors/recycling-waste. Accessed 30 Apr 2017
36. OECD (2016) Nanomaterials in waste streams. Current Knowledge on risks and impacts, organisation for economic cooperation and development
37. Organisation for Economic Cooperation and Development (OECD) (2012) Guidance on sample preparation and dosimetry for the safety testing of manufactured nanomaterials. Technical report 36. Organisation for Economic Co-operation and Development (OECD)
38. Rauscher H, Roebben G, Sanfeliu AB, Emons H, Gibson N, Koeber R, Linsinger T, Rasmussen K, Sintes JR, Sokull Klüttgen B, Stamm H (2015) Towards a review of the EC recommendation for a definition of the term "nanomaterial". Technical report. Joint Research Centre (JRC)
39. Rauscher H, Rasmussen K, Sokull Klüttgen B (2017) Regulatory Aspects of Nanomaterials in the EU. Chemie Ingenieur Technik 89(3):224–231
40. Ruiz ON, Fernando KAS, Wang B, Brown NA, Luo PG, McNamara ND, Vangsness M, Sun YP, Bunker CE (2011) Graphene oxide: a nonspecific enhancer of cellular growth. ACS Nano 5(10):8100–8107
41. Srinivas G, Burress J, Yildirim T (2012) Graphene oxide derived carbons (GODCs): synthesis and gas adsorption properties. Energy Environ Sci 5(4):6453–6459
42. The British Standards Institution (2012) Disposal of manufacturing process waste containing manufactured nano-objects. Guide. Technical report PAS 138:2012. British Standards Institution (BSI)

43. Upadhyay RK, Soin N, Roy SS (2014) Role of graphene/metal oxide composites as photocatalysts, adsorbents and disinfectants in water treatment: a review. RSC Adv 4(8):3823–3851

44. US-FDA (2014) Guidance for industry considering whether an FDA-regulated product involves the application of nanotechnology. Technical report. US Food and Drug Administration

45. Wang K, Ruan J, Song H, Zhang J, Wo Y, Guo S, Cui D (2010) Biocompatibility of graphene oxide. Nanoscale Res Lett 31(1):2302

Chapter 4
Remediation Process by Graphene Oxide

Abstract In this chapter the remediation of water by means of graphene oxide is reported. The advantage of the use of this material is based on the amphiphilic character of graphene oxide which allows to remove inorganic and organic molecules at nanoscale range. In a remediation process, graphene oxide can be implemented as free-standing, functionalized or composite material, as well as, in solid or liquid phase. Lamellar or foam arrangement and graphene oxide dispersion are differently used to produce membrane or 3D structures to efficiently remove a broad range of natural and synthetic molecules/ions with high sensitivity. Thus, graphene oxide opens an unconventional scenario for nanopurification of industrial wastewater which are contaminated with hazardous molecules. This subject is a relatively new topic and we are witnessing its development as a feasible nano-adsorbent/removal material. The most updated progress in this field is reported and categorized.

Keywords Remediation · Removal contaminants · GO Membranes · GO Liquid Phase · Functionalization · Hybrid Materials

The environmental remediation process is the act of removing pollutions from environmental media, such as water or soil [15, 29, 33]. This removal process is often categorized in "ex-situ" and "in-situ" approach which indicates that the extraction of contaminants occurs with and without extracting of media, i.e. soil or water, respectively. Examples of remediation technologies are Surfactant Enhanced Aquifer Remediation (SEAR), Pump and Treat, Soil Vapour Extraction (SVE). One of the best methods for removing pollutions consists of adsorbing the contaminant in a vector substance. A number of substrates are currently employed in the environmental remediation and they range from a conventional activated carbon to clays, various porous materials, surfactants and polymers [15, 20, 38]. Recently, the introduction of nanomaterials opens a new scenario inside the environmental remediation topic due to the advanced engineering properties of these smart materials.

© The Author(s) 2017   35
F. Pendolino and N. Armata, *Graphene Oxide in Environmental Remediation Process*, SpringerBriefs in Applied Sciences and Technology, DOI 10.1007/978-3-319-60429-9_4

4.1 Nanomaterials—Remediation Process

In the last decade, the use of nanomaterials in technology is increased exponentially ascribed to the non conventional properties. Several materials have found new life when a nanoform is employed in a remediation process. The use of nanotechnology in the field of remediation has led to the coining of a new term "nanoremediaton". For instance, TiO_2 nanowires reveal an appreciable photocatalytic effect when comparing with the conventional bulk material due to effects on the quantum scale and the higher surface area. The TiO_2 nanowires membranes include photocatalytic degradation, disinfection and filtration [15]. Recently, a novel environmental remediation approach is the nanoscale zero-valent iron (nZVI), also commercially available, are very effective against various pollutants such as pesticides, atmospheric toxic gases and other organic solvents [23]. The mechanism of action of nZVI exploits the reducing properties of the Fe^0 metal and the consequent degradation of the contaminant involved in the redox reaction. From a practical point of view, the nZVI can be included in Permeable Reactive Barriers (PRBS), technology that allows the passive removal of contaminants in situ, or they can be injected in the colloidal form in porous media. Moreover, nanoparticle colloidal dispersions show an alternative method to separate pollutions through adsorption and subsequent magnetic separation. In this method, recycling and reuse of nanoparticles is an advantage but the cost-effective limits industrial employment [15]. Among carbon based nanomaterials, fullerenes, nanotube, and other carbon allotropes are promising candidates for remediation process because of its peculiar structure. These nanomaterials are often used as sorbents, however, their photocatalytic, catalytic effects, as well as the low cost, should be considered as additional value in a remediation process. Despite these technologies, the bidimentional graphene or its oxidized form is coming to be employed in removal contaminants from the aqueous medium.

4.2 Graphene Oxide—Remediation Process

The contaminant capability of GO is correlated with several structural factors. The number of oxygen domains, the size of basal graphene plane and the re-arrangement of GO layers, plays a crucial role in removing pollutions. An advantage of using GO in remediation consists of several order of magnitude in the removal capability in comparison with conventional adsorbent materials [2, 5]. The amphiphilic behaviour of graphene oxide allows an ideal substrate for broad classes of molecules to be bonded or adsorbed on its structure. Hence, metal ions, synthetic or natural organic molecules, pharmaceuticals, agriculture molecules, biomolecules (proteins, DNA, etc.), or mixtures (oil/petrol) are potentially removed by graphene oxide. By that, the study and, consequently, the use of GO in removing contaminants is at its early stage. The majority of works on remediation using GO have been done under controlled laboratory conditions and the effect of a combination of factors (pH, ionic strength,

Fig. 4.1 Scheme illustrates
the main routes of using
graphene oxide in
environmental remediation:
free-standing, GO/Inert,
Hybrid GO and
Functionalized GO

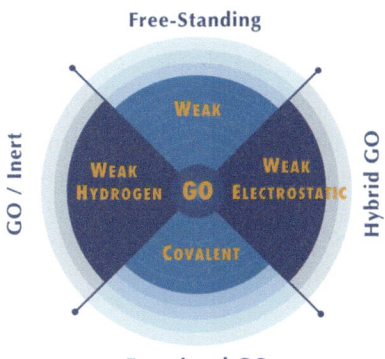

salts, etc.) is not fully yet understood. Therefore, the evaluation of a real contaminated medium needs further investigation before graphene oxide is going into the conventional remediation market. A number of reviews are suggested by authors for an in-deep screening of a variety of carbon/graphene based materials (graphene, nanotube, reduced graphene oxide, etc.) to be used in remediation [4, 5, 9, 12, 25, 27, 40, 45, 57].

In the follow paragraphs, we collected the most valuable and updated works on remediation process using graphene oxide as a liquid or solid phase. The scheme reported in Fig. 4.1 shows the major routes of using graphene oxide in remediation activities. Four principal approaches are identified: (*i*) pristine GO (free-standing); (*ii*) mixture of GO and inert material to operate as a support (GO/Inert); (*iii*) composite in which GO interacts with other materials to create a framework possessing advanced properties (Hybrid GO); (*iv*) functionalization of GO with a molecule that can interact through hydroxyl, epoxy and alkoxy groups presents on graphene basal plane of graphene oxide. The interactions of GO with other materials ranges from weak to covalent bonds allowing the possibility to modulate the physical chemical properties of graphene oxide depending on the applications.

4.2.1 Solid Phase GO

The solid phase of graphene oxide can be designed by two main forms, i.e. membrane and foam (sponge). A membrane is simply a barrier which allows the crossing of certain species, while blocking other based on the properties of the membrane and species. On the other hand, a foam is a three-dimentionals structure possessing a porous matrix with a defined size in which species are passed through or rejected. Depending on how the water GO dispersion is dried, GO layers can be arranged differently. By drying in an oven, a layer by layer aligns of GO basal graphene planes represent the most probable arrangement, thus a laminar and flexible materials are produced. On the other hand, when GO dispersion is dried by means of unidirectional freeze-drying method a 3D structure (sponge/foam) is made. Both forms are used as adsorbing materials thanks to their structure and, predominantly, to the presence of oxygen groups on graphene oxide surface. Besides, the strength of the GO structure appears flexible for lamellar form and fragile for the sponge form. Regardless, the properties of GO form strongly depended on a number of factors, such as the synthesis, oxygen content, presence of water or ions, and the treatment post synthesis (sonication, thermal dry, unidirectional freeze-drying, etc.) For this reasons, graphene oxide is often mixed with inert material (polymer, MOF, etc.) to improve the mechanical properties.

4.2.1.1 GO Membranes

A pristine GO membrane ends in a dispersion when treated in a water medium. However, the presence of cations reinforces the stability of GO membranes. Yeh et al. [52] demonstrated how the addition of Al^{3+}, Mn^{2+} or other multivalent metal cations allows a cross-linking with cations and strengthen the membrane. To over come to this issue, a typical GO membrane is made of a polymeric matrix or other support which improves the rigidity of the membrane. The development of membranes based on GO is in continuous progress.

Metal Ions. GO denotes a significant removal capability concerning the remediation of heavy metal ions (Cd^{2+}, Pd^{2+}, Hg^{2+}, Cr^{6+}, As^{3+}, etc.). Because of the positive charges, these ions interact with oxygen and negative functional groups on GO enhancing the cations capture. Tan et al. [44] reported a successful adsorption of Cu^{2+}, Cd^{2+} and Ni^{2+} on a GO/PVA membrane. The equilibrium was reached in a short time and the membrane was regenerated more than six times. Zhang et al. [54] investigated a composite membrane for nanofiltration made of GO deposited onto Torlon support via a layer-by-layer (LbL) approach. The membrane successfully removes more than 95% towards Ni^{2+}, Pd^{2+}, Zn^{2+} from aqueous systems and exhibits a long-term stability during a 150-h NF test.

Salts. Ions penetration through a pristine GO membrane (often named free-standing GO membrane) was investigated by Sun et al. [41] in 2011. Their founds showed that in a mixed solution, sodium salt (NaCl, NaOH, $NaHSO_3$, $NaHCO_3$) permeated fast through the GO membranes rather than heavy metal salts ($CuSO_4$, $CdSO_4$, $MnSO_4$). On the other hand, larger ions (Cu^{2+}) and organic molecules (Rhodamine B) were blocked by the membrane. Figure 4.2 shows the variation of permeation with time in the form of change in conductivity. Authors claim that the nanocapillaries within the lamellar structure and the oxygen functional groups are the responsible of the different ion penetration. Recently, Wang et al. [46] combine the properties of a MOF with GO to explore the efficiency of a desalination membrane. Authors chose a common zeolitic MOF, zeolitic imidazolate framework-8 (ZIF-8) which was grown onto graphene oxide surface. This composite MOF-GO material was able to remove four types of salt solutions, i.e. MgCl2, MgSO4, NaCl and Na2SO4. For a better

Fig. 4.2 **a** Penetration process of CuSO4 and NaCl mixture and **b** Rhodamine B and NaCl during 3 h of penetration. Each component has a concentration of 0.1 mol/L. The inset shows digital photos of the mixture and its filtrates. Reprinted with permission from [41]. Copyright (2012) American Chemical Society

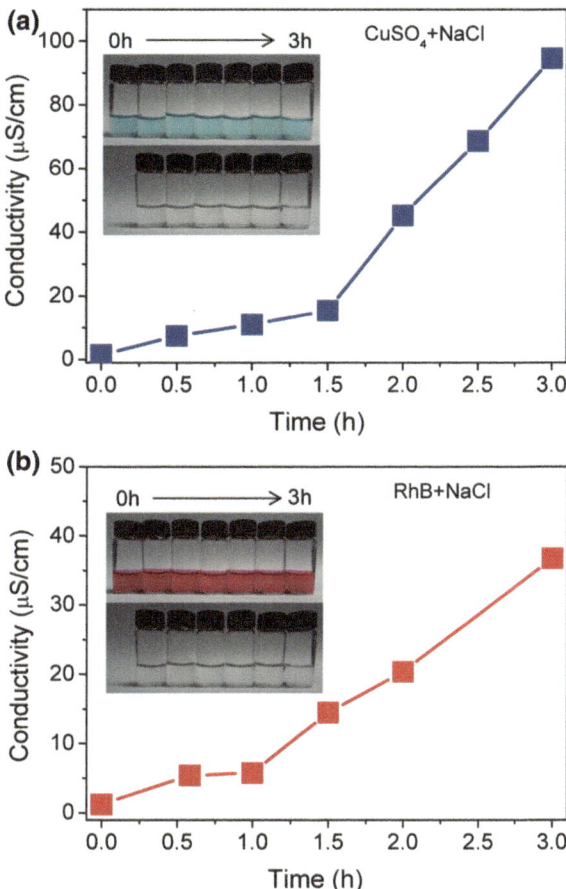

performance, ZIF-8/GO was immobilized using a PES membrane via interfacial polymerization. In addition the membrane exhibits an antimicrobial activity.

Organic Molecules. Aba et al. [1] reported a composite membrane GO on ceramic hollow fiber substrate. This GO-Ceramic membrane shows high performance using a flux of acetone or methanol and is promptly of rejects molecules larger than 300 Dalton. By using this membrane, a nanofiltration is practicable for a wide range of organic molecules.

Joshi et al. [24] studied micrometer-thick GO membranes to demonstrate the permeation of a feed in compartments filled with different liquids, including water, glycerol, toluene, ethanol, benzene, and dimethyl sulfoxide over a period of many weeks. A U-shaped tube is set up where GO membrane is placed in between the compartments. They found that small species permeate close to the same speed, whereas larger ions/molecules are not permeated. Finally, a cutoff of 4.5 Å was identified for species to be sieved out. In the same study, the influence of ions (MgCl$_2$,

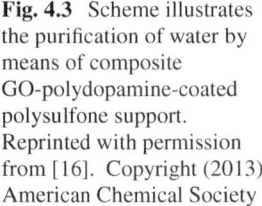

Fig. 4.3 Scheme illustrates the purification of water by means of composite GO-polydopamine-coated polysulfone support. Reprinted with permission from [16]. Copyright (2013) American Chemical Society

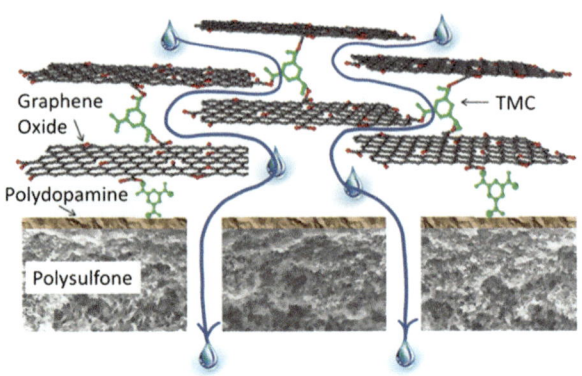

KCl, and $CuCl_2$) in their hydrated form was investigated. Authors argue that the permeation depends on size without being affected by the charge.

A non-selective GO membrane was prepared via spray or spin-coating on copper substrate by Nair et al. [35]. This membrane was able to selectively reject all molecules except H_2O. Authors tested, by weight, ethanol, hexane, acetone, decane, and propanol during several days, without any visible changes. This kind of membrane can be used for a clear nanofiltration of pure water.

Dyes. Hu et al. [16] synthesized a separation membrane via layer-by-layer deposition of GO nanosheets which was previously cross-linked by 1,3,5-benzenetricarbonyl group on a polydopamine-coated polysulfone support (Fig. 4.3). They found a flux ranged of 80–276 LMH/MPa that represents values much higher than commercial nanofiltration membranes. This membrane was capable of high rejection (95%) of Rhodamine-WT and moderate rejection (66%) of Methylene blue dyes.

Next, the Evans Blue dye was studied by Huang et al. [17] to investigate the filtration properties of GO nanochannels. They claimed that some molecules are blocked by altering the nanochannel network in membrane. Authors found 85% of the rejection rate for the dye and water flux of 71 L m^{-2} h^{-1} bar^{-1} by tuning pH, salt concentration and pressure. In a second publication, Huang et al. [18] observed ultra fast water flow through nanochannels possessing a sharp size distribution of 3–5 nm. Authors have studied a range of increasing size of molecules, as small as $[Fe(CN)_6]^{3-}$ (1 nm) up to Cytochrome C ($2.5 \times 2.5 \times 3.7$). Only gold nanoparticles (5 nm) had the 100% rejection and slightly less for Cytochrome C.

Oil-in-water Emulsion. An advanced engineering a sol-gel GO membrane was proposed by Huang et al. [19] to "separate a series of surfactant-free and surfactant-stabilized oil-in-water emulsion driven solely by gravity". The sol-gel GO membrane was cross-linked with a branched polyethyleneimine (PEI) via a one-step process and, then, heated to fabricate a 3D network structure. Proposed non-laminated sol-gel GO membrane exhibits superhydrophilic and underwater superoleophobic properties which permit to separate surfactant-free and surfactant stabilized undissolved oil-in-water emulsions. By only gravity, the efficient separation was close to 99%.

4.2.1.2 GO Foam (Sponge)

Metal Ions. Mi et al. [34] reported a kinetic study of removing Cu^{2+} from water using a GO aerogel made by means of unidirectional freeze-drying method. Authors found a pseudo second-order kinetic model with a dependence from the concentration of Cu^{2+} and an equilibrium which is reached in about 15 min (shorter than CNT, for 24 h). Moreover, the Cu^{2+} adsorption resulted as pH dependent and the removal efficiency increased from 32.3 to 96.8% when the pH changes from 2 to 6. Henriques et al. [13] showed the enhanced removal capacity of a modified GO foam against mercury cation. Authors compared a pristine GO foam and a functionalization with nitrogen (N) and sulfur (S) groups. The best removal capacity was found in GO foam functionalized with N that was able to adsorb 96% of Hg^{2+} within 24 h, leading to a limit value suggested by normative.

Dyes. A hydrogel composite of GO 3D architecture and ssDNA was used by Xu et al. [49] to remove safranie O dye. After 24 h the 100% (960 mg × g^{1-}) of the dye was adsorbed into GO-ssDNA (Fig. 4.4). Authors claim that strong electrostatic interactions are made between the positive charge present on safranie O and the

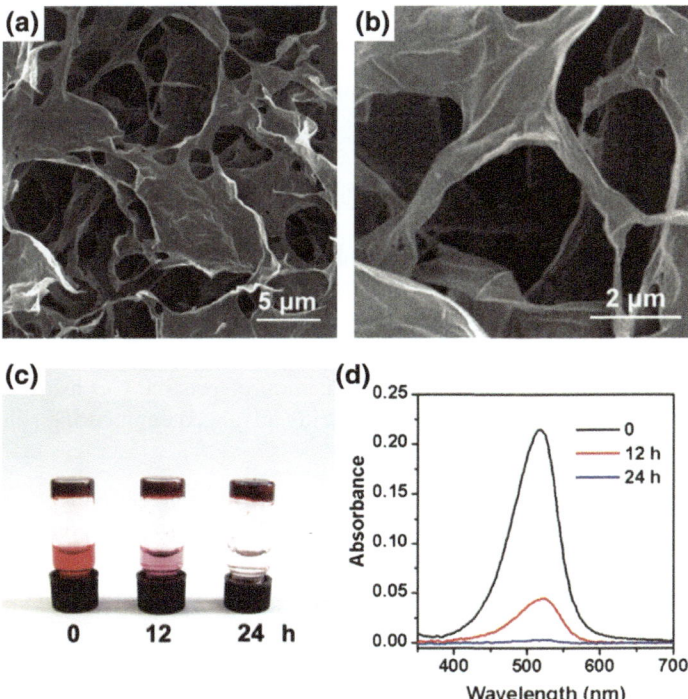

Fig. 4.4 a–b SEM images of GO/DNA microstructures. **c–d** water dispersion of GO/DNA composite and safranine O at different times. Reprinted with permission from [49]. Copyright (2010) American Chemical Society

negative charges of GO and DNA. A 3D structure of GO was adopted by Li et al. [30] to remove water soluble dyes. Authors showed how this form of GO is able to remove methylene blue and methyl violet from water. The process was very fast reaching the equilibrium in about 2 min and an adsorption capacity of 397 and 467 mg × g^{1-}, respectively. In practice, the 99.1% of methylene blue and 98.8% of methyl violet have been removed. Authors are justified the high adsorption of organic molecules by means of strong $\pi - \pi$ interaction on the GO surface.

4.2.1.3 GO—Gas Removal

The use of GO as a gas sensor can be viewed, on the other side, as an adsorbent material at ppm range. The presence of chemical groups on GO helps interaction with a gas molecule such as NO_2 or humidity. Choi et al. [7] reported that pristine GO film is active towards NO_2, NH_3 and H_2 via C-O bonds up to 500 ppm concentrations. Bi et al. [3] showed that a GO film inserted into a circuit is able to adsorb humidity and was modulated by different frequencies. A 3D GO structure is showed to be sensible towards volatile organic compound (VOC) by Matsuyama et al. [32]. The authors have used acetone or 1-butanol as VOC prototypes. 3D graphene oxide device is responsive to a broad concentration (up to 1000 pm) range for VOC.

4.2.2   Liquid Phase GO

Despite solid GO is the more convenient form to be used in environmental remediation, the liquid phase represents an alternative way to purify water. The advantage of GO dispersion resides in utilizing a chemical interactions between the functional groups of GO and a target contaminant. In addition, both faces of GO layer are efficiently employed to create weak or chemical bonds. The presence of hydroxyl, epoxy and alkoxy groups on GO allows many of conventional organic synthesises to add an auxiliary functionality. The functionalization of pristine GO leads to modulate the physical chemical properties for producing advanced engineering nanomaterials which help an in-deep pollutant removal.

Metal Ions. Pristine GO membranes were employed for removing several cations from aqueous solution. Rahimi et al. [37] investigated the capacity of adsorbing for Cd^{2+}, Pb^{2+}, Zn^{2+}, Mn^{2+}, Cu^{2+}, Fe^{3+} by pristine GO. The physical chemical sorption revealed the high efficiency of pristine GO in removing metal ions. The increasing of pH influenced the removal efficiency positively. The systems obeyed to a pseudo-second-order kinetics and Langmuir models. Zhao et al. [56] reported a removal capacity of 842 mg × g^{1-} (293 K), 1150 mg × g^{1-} (313 K), and 1850 mg × g^{-1} (333 K) for lead (Pb^{2+}). In another study [55], authors reported the adsorption of Cd^{2+} and Co^{2+} from GO dispersion. A constant temperature 303 K and pH 6.0, it was found removal capacity of 106.3 and 68.2 mg × g^{-1}, respectively. Moreover, lanthanides

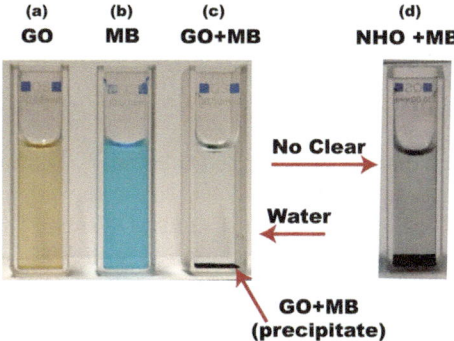

Fig. 4.5 Digital photos of **a** GO dispersion (0.05 mg/ml); textbfb methylene *blue* solution (5 mg/ml); textbfc mixture of GO and MB, the supernatant results in a clear liquid phase and a precipitate composed by GO+MB; textbfc dispersion of carbon nanohorn oxide and MB, in the supernatant a residual nanohorn oxide and MB is present

cations can also be removed by means of GO membranes. Sun et al. [42] studied the interaction of GO with Eu(III) resulting in an adsorption of 175.44 mg $\times$ g^{1-}, while, Pan et al. [36] removed from a dispersion 98.7% of Th^{4+} in about 10 min. Xu et al. [48] reported the interaction between GO and Th^{4+} as a function of contact time, pH, ionic strength, GO concentration and temperature. They obtained as maximum adsorption capacity of 252.5 μmol/g. Interaction with U^{6+} was investigated by Li et al. [28] on 2012. The variation of ionic strength is taken into account, initial concentration of U^{6+} and pH, and evaluates the adsorption of 299 mg $\times$ g^{1-} at pH 4. GO was also employed to separate Th^{4+}/U^{6+} cations mixture, as described by Jiang et al. [22]. In their study GO showed a great selectivity for Th^{4+} and the separation factor of Th^{4+}/U^{6+} was as high as 36 after 2 h of contact at pH 3.8.

Dyes. A typical example of organic dye removal using liquid GO dispersion involves a prototype contaminant, methylene blue dye (MB). As seen in Fig. 4.5, the cuvette (a) contains a GO dispersion (0.05 mg/ml, pH 7) and the cuvette (b) a methylene blue solution (5 mg/ml), while in cuvette (c) is shown the precipitate (GO+MB) after decanting time and leaving a clear supernatant. The result exhibits a removal contaminant of 95% with GO having a concentration of 100 times lower than that of the dye. The efficiency of GO was also compared with other oxidized carbon allotropes disperse in the liquid phase, i.e. carbon nanohorn oxide. In Fig. 4.5 cuvette (d), the contaminant dispersion of nanohorn oxide results in a non clear liquid phase, demonstrating that in this case the efficiency was very low, even after 3 days of decanting. This undoubtedly demonstrates the much greater removal efficiency of GO. Similar result was obtained by Yang et al. [51] to purify waster system from methylene blue via GO adsorption. Authors reported a removal efficiency of 99% for a concentration of MB below 200 mg/L and a maximum absorption capacity of 714 mg/g. This little difference in removal efficiency between Yang results (99%) and the example above (95%) depends on the concentration ratio MB/GO. In the first case, the maximum ratio is 0.6 with an oxygen content in GO of about 50% (Hummers method), on the opposite, the example above has a MB/GO ratio of 100 and an oxygen content in GO of about 20% (4-steps method). Thus, the synthetic method and, consequently, the GO structure arises to a distinct removal efficiency.

Antibiotic. Emerging contaminants are represented by antibiotics which are frequently detected in wastewater or water bodies. An example of remediation by means of GO is reported by Dong et al. [8]. Authors studied the removal of Levofloxacin and lead (Pb) solutions in a dispersion of graphene oxide both as a single component or mixture. Langmuir maximum adsorption capacities of 256.6 and 227.2 mg $\times$ g$_{-1}$ for Levofloxacin and Pb, respectively. They claim that GO is a promising nanoadsorbent with high- efficiency for water treatment.

4.2.3 Functionalized GO

The overall removal capacities of graphene oxide could be improved by functionalizing chemical groups on graphene oxide. Theoretically, any molecules can be bonded and extend the physical chemical properties [40] through interactions with oxygen domains, such as hydroxyl, carbonyl and alkoxy.

EDTA-GO

Removal performance respect to the Pb^{2+} has improved using a functionalized GO substrate with ethylenediamine triacetic acid (EDTA), as showed by Madadrang et al. [31]. The mechanism adsorption is illustrated in Fig. 4.6. EDTA was previously silanized with N-(trimethoxysilylpropyl) and subsequently added to the hydroxyl groups on GO. The adsorption isotherms showed a higher performance for the EDTA-GO composite in comparison with the pristine GO and a maximum adsorption capacity for Pb^{2+} of 479 mg/g. The advantage of regenerate the EDTA-GO by washing with HCl suggests a convenient composite material to remove toxic metal cations from aqueous systems.

Chitosan-GO

Recently, the functionalization of GO with Chitosan appears significantly adapted to remove a broad classes of contaminants. He et al. [11] reported as the porous Chitosan-GO material increased the adsorption capacity of Pb^{2+} up to 99 mg/g when 5 wt% of GO is added. Wang et al. [47] also reported the removal efficiency of aerogel Chitosan-GO towards azo dyes, i.e. methyl orange and amido black 10B. They found adsorption capacity of 686.89 mg/g and 573.47 mg/g, respectively. Moreover, Yan et al. [50] performed measurements on the same system by removing simultaneously dyes and metal ions. Precisely, methylene blue/Cu^{2+} and methyl orange/Cr^{4+} systems were analyzed. These systems reached equilibrium within 10 min and exhibit a regeneration performance. A step forward was done by a further modification of Chitosan-GO system by adding magnetic nanoparticles. An example is reported by

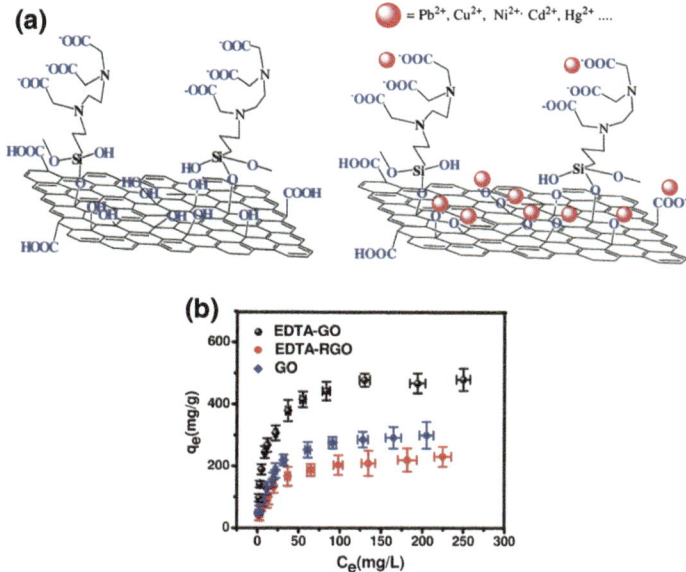

Fig. 4.6 a Chemical structure of functionalized GO with EDTA and its interaction with heavy metal cations. **b** Adsorption isotherms of Pb (II) on GO, EDTA-rGO and EDTA-rGO at pH 6.8. Reprinted with permission from [31]. Copyright (2012) American Chemical Society

Gul et al. [10]. Authors studied the removal efficiency of Fe3O4/Chitosan-GO composite towards textile dyes, i.e. methyl violet and Alizarin yellow R. Cationic and anionic dyes were removed from water with higher efficiency than pristine chitosan or GO.

Sulfonyl-GO

Newly, removal capability of U^{6+} was improved by the modification of GO using sulfonyl ($-OSO3H$) groups. Sun et al. [43] demonstrated that solfonated GO is able to adsorb 45.05 mg/g at pH 2.0 of U^{6+} solution and the sorption is independent from the ionic strength. Authors are claimed that sulfonyl groups are responsible for highly effective adsorption at so low pH conditions.

Amine-GO

Hosseinabadi-Farahani et al. [14] have modified the GO surface with 3-aminopropyl-trimethoxysilane in a way that an amine group is active to interact with a target dye. Anionic dyes were chosen: Acid Blue 92 (AB92), Acid Red 14 (AR14), and Direct Green 6 (DG6) and an adsorption capacity of 60, 64 and 65 mg/g, respectively, were calculated by Langmuir model.

TiO$_2$-GO

Zhang et al. [53] reported the modification of the GO structure by means of UV irradiation to get additional carboxylic groups and subsequent esterification with

TiO_2. The functionalized GO materials were employed for a photocatalytic reduction of Cr(VI) to Cr(III) when a graphene base materials were loaded at 5% of TiO_2.

Ceria-GO

A functionalization of GO by CeO_2 was employed by Sakthivel et al. [39] to almost remove the 99.99% of arsenic ion from contaminated liquid phase. They measured adsorption capacity of 185 mg/g for As(III) and 212 mg/g for As(V). A comparable evaluation of Ceria$_2$-GO with other adsorbent materials, such as Fe3O4 graphene MnO2 composites or Hybrid (polymeric/inorganic), fibrous sorbent, set the GO composite as one of the best support possessing a high maximum adsorption capacity.

4.2.4 Hybrid Material M-GO

By following the advanced engineering of nanomaterial, the enhancement of environmental remediation behaviour can be reached by redesigning a material based on graphene oxide. This approach extends towards a composite that ranges from inorganic to polymeric up to magnetic framework.

MOF-GO

Further engineering material is represented by the composite made of metal-organic framework and graphene oxide (MOF/GO). This composite takes advantage of high porosity structure from MOF and the reactive centres of GO to capture molecules. Jabbari et al. [21] have synthesized via green method a MOF@GO composite to remove organic pollutants. Specifically, authors used copper benzene-1,3,5-tricarboxylate (Cu-BTC MOF) as a MOF crystal to be implemented into the GO structure. Furthermore, the addition of magnetic Fe_3O_4 nanoparticles enhances the adsorption capacity. Results showed that in the composites the removal ability was higher than the parent materials.

Magnetic-GO

A composite made of $Fe@Fe_2O_3/GO$ was employed by Li et al. [26] to extract bisphenol A (BPA), triclosan (TCS) and 2, 4-dichlorophenol (2, 4-DCP) from aqueous solution. These three molecules are endocrine disrupting phenols so toxic compounds which are part of the industrial production of plastic, coke, pesticide, etc. The electrostatic interaction between $Fe@Fe_2O_3$ shell and GO allows a separation by means of a magnetic field applied to the sample. This study demonstrates the nano extraction with a range from 0.08 to 0.10 ng mL^{1-} having an extraction time of about 10 min (Fig. 4.7).

Polymers-GO

Macroporous polystyrene microsphere/graphene oxide (PS/GO) composite monolith was investigated by Chen at al. [6] as an adsorbent material for tetracycline in aqueous solution. Authors claim an exceptional mechanical strength for the three-dimensional macroporous structure with a size of ranging from 4 to 20 µm. The adsorption

Fig. 4.7 a Scheme illustrates the preparation of composite Fe@Fe2O3/GO. **b** Removal efficiency (recovery %) for the parent systems, Fe@Fe2O3 and GO, in comparison with Fe@Fe2O3/GO for Bisphenol A (BPA), triclosan (TCS) and 2,4-dichlorophenol (2,4-DCP). Reprinted from Publication [26]. Copyright (2015), with permission from Springer

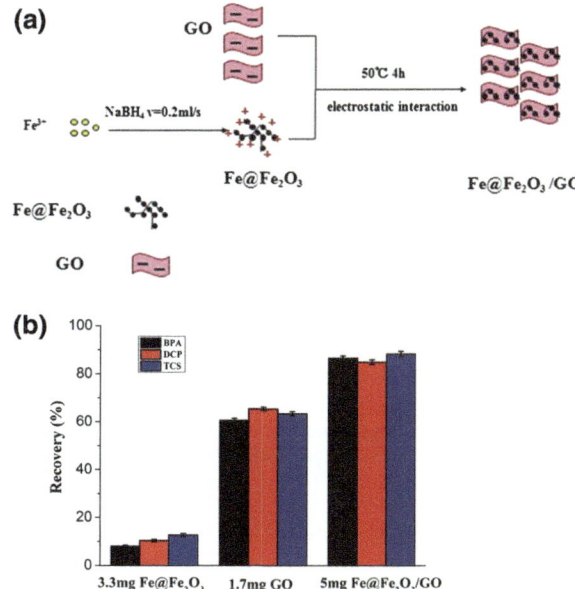

capacity was studied by UV-vis technique obtaining an ability 1.4 times higher for the composite comparing with the pristine GO. The maximum adsorption capacity was measured in $197.9\,\mathrm{mg} \times \mathrm{g}^{1-}$ at pH 6.

References

1. Aba NFD, Chong JY, Wang B, Mattevi C, Li K (2015) Graphene oxide membranes on ceramic hollow fibers—microstructural stability and nanofiltration performance. J Membr Sci 484:87–94

2. Bi H, Xie X, Yin K, Zhou Y, Wan S, He L, Xu F, Banhart F, Sun L, Ruoff RS (2012) Spongy graphene as a highly efficient and recyclable sorbent for oils and organic solvents. Adv Funct Mater 22(21):4421–4425

3. Bi H, Yin K, Xie X, Ji J, Wan S, Sun L, Terrones M, Dresselhaus MS (2013) Ultrahigh humidity sensitivity of graphene oxide. Sci Rep 3:2714 EP

4. Chabot V, Higgins D, Yu A, Xiao X, Chen Z, Zhang J (2014) A review of graphene and graphene oxide sponge: material synthesis and applications to energy and the environment. Energy Environ Sci 7(5):1564

5. Chen B, Ma Q, Tan C, Lim TT, Huang L, Zhang H (2015) Carbon-based sorbents with three-dimensional architectures for water remediation. Small 11(27):3319–3336

6. Chen LC, Lei S, Wang MZ, Yang J, Ge XW (2016) Fabrication of macroporous polystyrene/graphene oxide composite monolith and its adsorption property for tetracycline. Chin Chem Lett 27(4):511–517

7. Choi YR, Yoon YG, Choi KS, Kang JH, Shim YS, Kim YH, Chang HJ, Lee JH, Park CR, Kim SY, Jang HW (2015) Role of oxygen functional groups in graphene oxide for reversible room-temperature NO2 sensing. Carbon 91:178–187

8. Dong S, Sun Y, Wu J, Wu B, Creamer AE, Gao B (2016) Graphene oxide as filter media to remove levofloxacin and lead from aqueous solution. Chemosphere 150:759–764

9. Georgakilas V, Tiwari JN, Kemp KC, Perman JA, Bourlinos AB, Kim KS, Zboril R (2016) Non-covalent functionalization of graphene and graphene oxide for energy materials, biosensing, catalytic, and biomedical applications. Chem Rev 116(9):5464–5519

10. Gul K, Sohni S, Waqar M, Ahmad F, Norulaini NAN, AK MO (2016) Functionalization of magnetic chitosan with graphene oxide for removal of cationic and anionic dyes from aqueous solution. Carbohydr Polym 152:520–531

11. He YQ, Zhang NN, Wang XD (2011) Adsorption of graphene oxide/chitosan porous materials for metal ions. Chin Chem Lett 22(7):859–862

12. Hegab HM, Zou L (2015) Graphene oxide-assisted membranes: fabrication and potential applications in desalination and water purification. J Membr Sci 484:95–106

13. Henriques B, Gonçalves G, Emami N, Pereira E, Vila M, Marques PAAP (2016) Optimized graphene oxide foam with enhanced performance and high selectivity for mercury removal from water. J Hazard Mater 301:453–461

14. Hosseinabadi-Farahani Z, Mahmoodi NM, Hosseini-Monfared H (2015) Preparation of surface functionalized graphene oxide nanosheet and its multicomponent dye removal ability from wastewater. Fibers Polym 16(5):1035–1047

15. Hu A, Apblett A (eds) (2014) Nanotechnology for water treatment and purification. Springer, Heidelberg

16. Hu M, Mi B (2013) Enabling graphene oxide nanosheets as water separation membranes. Environ Sci Technol 47(8):3715–3723

17. Huang H, Mao Y, Ying Y, Liu Y, Sun L, Peng X (2013a) Salt concentration, pH and pressure controlled separation of small molecules through lamellar graphene oxide membranes. Chem Commun 49(53):5963–5965

18. Huang H, Song Z, Wei N, Shi L, Mao Y, Ying Y, Sun L, Xu Z, Peng X (2013b) Ultrafast viscous water flow through nanostrand-channelled graphene oxide membranes. Nat Comms 4:2979

19. Huang T, Zhang L, Chen H, Gao C (2015) Sol–gel fabrication of a non-laminated graphene oxide membrane for oil/water separation. J Mater Chem A 3(38):19517–19524

20. Ismadji S, Soetaredjo FE, Ayucitra A (2015) Clay materials for environmental remediation. Springer Briefs in Molecular Science. Springer, Heidelberg

21. Jabbari V, Veleta JM, Zarei-Chaleshtori M, Gardea-Torresdey J, Villagrán D (2016) Green synthesis of magnetic MOF@GO and MOF@CNT hybrid nanocomposites with high adsorption capacity towards organic pollutants. Chem Eng J 304:774–783

22. Jiang D, Liu L, Pan N, Yang F, Li S, Wang R, Wyman IW, Jin Y, Xia C (2015) The separation of Th(IV)/U(VI) via selective complexation with graphene oxide. Chem Eng J 271:147–154

23. Joo SH, Cheng F (2006) Nanotechnology for environmental remediation. Modern inorganic chemistry. Springer Science and Business Media, New York

24. Joshi RK, Carbone P, Wang FC, Kravets VG, Su Y, Grigorieva IV, Wu HA, Geim AK, Nair RR (2014) Precise and ultrafast molecular sieving through graphene oxide membranes. Science 343(6172):752–754

25. Joshi RK, Alwarappan S, Yoshimura M, Sahajwalla V, Nishina Y (2015) Graphene oxide: the new membrane material. Appl Mater Today 1(1):1–12

26. Li F, Cai C, Cheng J, Zhou H, Ding K, Zhang L (2015a) Extraction of endocrine disrupting phenols with iron-ferric oxide core-shell nanowires on graphene oxide nanosheets, followed by their determination by HPLC. Microchim Acta 182(15–16):2503–2511

27. Li F, Jiang X, Zhao J, Zhang S (2015b) Graphene oxide: a promising nanomaterial for energy and environmental applications. Nano Energy 16:488–515

28. Li Z, Chen F, Yuan L, Liu Y, Zhao Y, Chai Z, Shi W (2012) Uranium(VI) adsorption on graphene oxide nanosheets from aqueous solutions. Chem Eng J 210:539–546

29. Lichtfouse E, Schwarzbauer J, Robert D (eds) (2012) Environmental chemistry for a sustainable world. Springer, Netherlands

30. Liu F, Chung S, Oh G, Seo TS (2012) Three-dimensional graphene oxide nanostructure for fast and efficient water-soluble dye removal. ACS Appl Mater Interfaces 4(2):922–927

31. Madadrang CJ, Kim HY, Gao G, Wang N, Zhu J, Feng H, Gorring M, Kasner ML, Hou S (2012) Adsorption behavior of EDTA-graphene oxide for Pb (II) removal. ACS Appl Mater Interfaces 4(3):1186–1193
32. Matsuyama S, Sugiyama T, Ikoma T, Cross JS (2016) Fabrication of 3D graphene and 3D graphene oxide devices for sensing VOCs. MRS Adv 1(19):1359–1364
33. Meuser H (2012) Soil remediation and rehabilitation, treatment of contaminated and disturbed land, vol 23. Springer Science and Business Media, Dordrecht
34. Mi X, Huang G, Xie W, Wang W, Liu Y, Gao J (2012) Preparation of graphene oxide aerogel and its adsorption for Cu2+ ions. Carbon 50(13):4856–4864
35. Nair RR, Wu HA, Jayaram PN, Grigorieva IV, Geim AK (2012) Unimpeded permeation of water through helium-leak-tight graphene-based membranes. Science 335(6067):442–444
36. Pan N, Guan D, He T, Wang R, Wyman I, Jin Y, Xia C (2013) Removal of Th4+ ions from aqueous solutions by graphene oxide. J Radioanal Nucl Chem 298(3):1999–2008
37. Rahimi E, Mohaghegh N (2015) Removal of toxic metal ions from sungun acid rock drainage using mordenite zeolite, graphene nanosheets, and a novel metal-organic framework. Mine Water Environ 35(1):18–28
38. Ray SC (2015) Application and uses of graphene oxide and reduced graphene oxide. Elsevier
39. Sakthivel TS, Das S, Pratt CJ, Seal S (2017) One-pot synthesis of a ceria-graphene oxide composite for the efficient removal of arsenic species. Nanoscale 9(10):3367–3374
40. Smith SC, Rodrigues DF (2015) Carbon-based nanomaterials for removal of chemical and biological contaminants from water: a review of mechanisms and applications. Carbon 91:122–143
41. Sun P, Zhu M, Wang K, Zhong M, Wei J, Wu D, Xu Z, Zhu H (2012a) Selective ion penetration of graphene oxide membranes. ACS Nano 7(1):428–437
42. Sun Y, Wang Q, Chen C, Tan X, Wang X (2012b) Interaction between Eu(III) and graphene oxide nanosheets investigated by batch and extended X-ray absorption fine structure spectroscopy and by modeling techniques. Environ Sci Technol 46(11):6020–6027
43. Sun Y, Wang X, Ai Y, Yu Z, Huang W, Chen C, Hayat T, Alsaedi A, Wang X (2017) Interaction of sulfonated graphene oxide with U(VI) studied by spectroscopic analysis and theoretical calculations. Chem Eng J 310:292–299
44. Tan P, Sun J, Hu Y, Fang Z, Bi Q, Chen Y, Cheng J (2015) Adsorption of Cu2+, Cd2+ and Ni2+ from aqueous single metal solutions on graphene oxide membranes. J Hazard Mater 297:251–260
45. Upadhyay RK, Soin N, Roy SS (2014) Role of graphene/metal oxide composites as photocatalysts, adsorbents and disinfectants in water treatment: a review. RSC Adv 4(8):3823–3851
46. Wang J, Wang Y, Zhang Y, Uliana A, Zhu J, Liu J, Van der Bruggen B (2016) Zeolitic imidazolate framework/graphene oxide hybrid nanosheets functionalized thin film nanocomposite membrane for enhanced antimicrobial performance. ACS Appl Mater Interfaces 8(38):25508–25519
47. Wang Y, Xia G, Wu C, Sun J, Song R, Huang W (2015) Porous chitosan doped with graphene oxide as highly effective adsorbent for methyl orange and amido black 10B. Carbohydr Polym 115:686–693
48. Xu H, Li G, Li J, Chen C, Ren X (2016) Interaction of Th(IV) with graphene oxides: batch experiments, XPS investigation, and modeling. J Mol Liq 213:58–68
49. Xu Y, Wu Q, Sun Y, Bai H, Shi G (2010) Three-dimensional self-assembly of graphene oxide and DNA into multifunctional hydrogels. ACS Nano 4(12):7358–7362
50. Yan H, Yang H, Li A, Cheng R (2016) pH-tunable surface charge of chitosan/graphene oxide composite adsorbent for efficient removal of multiple pollutants from water. Chem Eng J 284:1397–1405
51. Yang ST, Chen S, Chang Y, Cao A, Liu Y, Wang H (2011) Removal of methylene blue from aqueous solution by graphene oxide. J Colloid Interface Sci 359(1):24–29
52. Yeh CN, Raidongia K, Shao J, Yang QH, Huang J (2015) On the origin of the stability of graphene oxide membranes in water. Nat Chem 7(2):166–170

53. Zhang K, Kemp KC, Chandra V (2012) Homogeneous anchoring of TiO2 nanoparticles on graphene sheets for waste water treatment. Mater Lett 81:127–130
54. Zhang Y, Zhang S, Gao J, Chung TS (2016) Layer-by-layer construction of graphene oxide (GO) framework composite membranes for highly efficient heavy metal removal. J Membr Sci 515:230–237
55. Zhao G, Li J, Ren X, Chen C, Wang X (2011a) Few-layered graphene oxide nanosheets as superior sorbents for heavy metal ion pollution management. Environ Sci Technol 45(24):10454–10462
56. Zhao G, Ren X, Gao X, Tan X, Li J, Chen C, Huang Y, Wang X (2011b) Removal of Pb(II) ions from aqueous solutions on few-layered graphene oxide nanosheets. Dalton Trans 40(41):10945–10952
57. Zhao J, Liu L, Li F (2015) Graphene oxide: physics and applications. Springer Briefs in Physics. Springer, Heidelberg

Chapter 5
Conclusion and Outlook

Abstract The graphene oxide is explored as a smart material in the context of environmental remediation. This material possesses a wide potential in removal pollutions from contaminated water systems. The advantage of graphene oxide consists of an affinity for inorganic and organic contaminants and placing graphene oxide in a predominant position in comparison with conventional adorbant/removal materials.

Keywords Graphene oxide · Removal contaminants · Pollution · Enviromental remediation

In the last decade, graphene oxide leads role either in the academic and industrial world due to its unconventional properties of the bidimentional structure in comparison with typical bulk materials. The presence of oxygen domains on the graphene basal plane interrupts the sp^2 network and allows the interactions with a target molecule. In addition, these groups provide for active sites for a further functionalization which enhances the properties of graphene oxide to engineering advanced materials.

In this context that the technological development is relevant for obtaining an adsorbent material with good efficiency to guarantee high quality water and preserve the natural freshwater sources. The conventional removal methods are still useful nowadays but overcoming this technology is providing a responsive technology for enhancing the quality of recycled water, reducing the level of contaminants to a level that is safe for use. The latest research shows that removal of chemical hazards using graphene oxide is a quite novel approach in the treatment of wastewater.

This smart material offers unconventional physical chemical properties that can work in removal technology restricting the negative impact to the environment and human health. The innovative features of the graphene oxide take into account several aspects in comparison with conventional technology. First, regular filters are based primarily on pore size molecular discrimination. Instead, the applicant wishes to implement a different way in which the removal of a molecule is originated from the reactivity of the removal materials. Thus, molecular control constitutes a very innovative aspect of graphene oxide material. Secondly, graphene oxide is capable of simultaneously removing of a broad class of molecules.

© The Author(s) 2017   51
F. Pendolino and N. Armata, *Graphene Oxide in Environmental Remediation Process*, SpringerBriefs in Applied Sciences and Technology, DOI 10.1007/978-3-319-60429-9_5

Although significant effort has been made to promote the graphene oxide within removal adsorption material, the market application, based on this technology, are still at the early stage with many remaining difficulties and challenges. The synthetic method is one of the issues to be accounted on controlling the final structure features, and the subsequent mass production. Next, this process has to evolve a green technology with less energy consumption. Finally, the availability of starting carbon sources which should be cheap to furnish a product with a reasonable market price. Moreover, the normative on graphene oxide and more in general for nanomaterials is not clear enough to promote this kind of materials in daily objects with the consequence of restraining its market.

In conclusion, the outstanding performance of the graphene oxide in environmental remediation raises hopes in the further development of industrial applications. More studies of fundamental science and collaboration between academia and industry will benefit the advancement of novel remediation technology based on carbon nanomaterial.